AFGHANISTAN
ARGENTINA
AUSTRALIA
AUSTRIA
BAHRAIN
BERMUDA
BOLIVIA
BOSNIA AND HERZOGOVINA
BRAZIL
CANADA
CHILE
CHINA
COSTA RICA
CROATIA
CUBA
EGYPT
ENGLAND
ETHIOPIA
FRANCE
REPUBLIC OF GEORGIA
GERMANY
GHANA
GUATEMALA
ICELAND
INDIA
INDONESIA
IRAN
IRAQ
IRELAND
ISRAEL
ITALY
JAMAICA
JAPAN
KAZAKHSTAN
KENYA
KUWAIT
MEXICO
THE NETHERLANDS
NEW ZEALAND
NIGERIA
NORTH KOREA
NORWAY
PAKISTAN
PERU
THE PHILIPPINES
RUSSIA
SAUDI ARABIA
SENEGAL
SCOTLAND
SOUTH AFRICA
SOUTH KOREA
TAIWAN
TURKEY
UKRAINE
UZBEKISTAN

China

Gary T. Whiteford
University of New Brunswick
Fredericton, New Brunswick

Series Consulting Editor
Charles F. Gritzner
South Dakota State University

Frontispiece: Flag of China

Cover: Workers in a rice paddy below the Karst Mountains

China

Chelsea House
An imprint of Infobase Publishing
132 West 31st Street
New York NY 10001

Library of Congress Cataloging-in-Publication Data
Whiteford, Gary T., 1941–
China / Gary T. Whiteford.—Updated ed.
p. cm.—(Modern world nations)
Includes bibliographical references and index.
ISBN 0-7910-8661-5
1. China—Juvenile literature. [1. China.] I. Title. II. Series.
DS706.W58 2005
951—dc22 2005041294

Chelsea House books are available at special discounts when purchased in bulk quantities for businesses, associations, institutions, or sales promotions. Please call our Special Sales Department in New York at (212) 967-8800 or (800) 322-8755.

You can find Chelsea House on the World Wide Web at http://www.chelseahouse.com

Cover and series design by Takeshi Takahashi

Printed in the United States of America

Bang 21C 10 9 8 7 6 5 4 3

This book is printed on acid-free paper.

Table of Contents

China

Acknowledgments

I would like to express my appreciation to the following people: My dear friend, Dr. Larry N. Shyu, Professor Emeritus Chinese History, University of New Brunswick, was a tremendous help. His intimate knowledge of China was invaluable in his guidance and review of every detail of this book. He was most giving of his time, advice, and patience. I am very grateful to him for his most gracious assistance. He was particularly helpful for comments on Chapter 3: China Through the Centuries. My colleague and friend, Professor Rodney H. Cooper, Computer Science and Computational Chemistry, University of New Brunswick, helped to carefully edit the chapters and meticulously checked all the figures and measurement conversions. Angela Wilkins faithfully typed and retyped the manuscript and was helpful in many other ways. The Executive Editor, Lee Marcott of Chelsea House Publishers, gave me encouragement throughout the entire project and helped bring about its completion. Dr. Charles F. Gritzner, Geographer, South Dakota State University and Series Consulting Editor, was most helpful with his comments, suggestions, and corrections. Last and by no means least, I owe an enormous debt of gratitude to my dear wife, Carole. This is her work, too.

Gary T. Whiteford
Fredericton, New Brunswick
January 2005

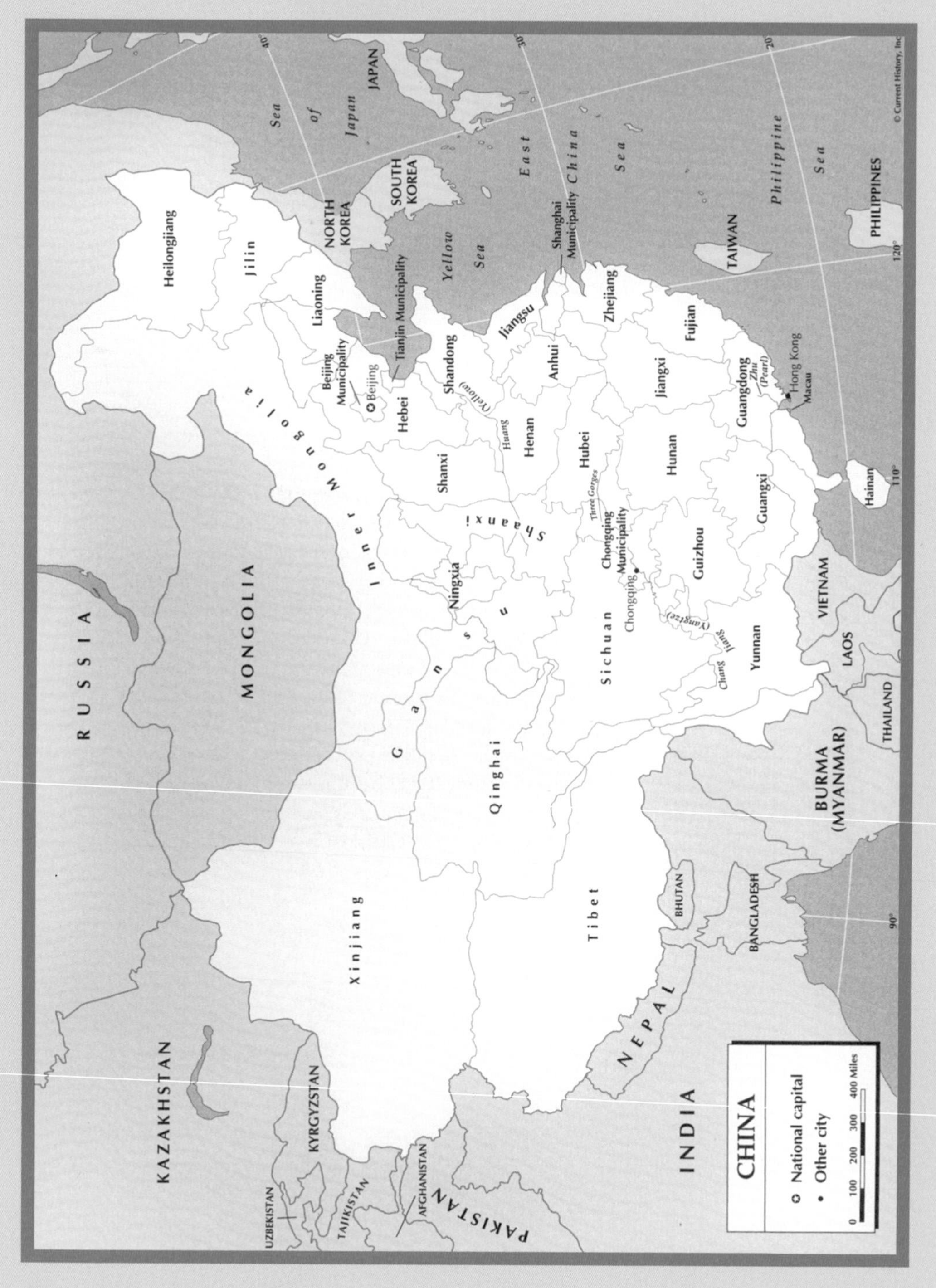

Map of the Provinces of China

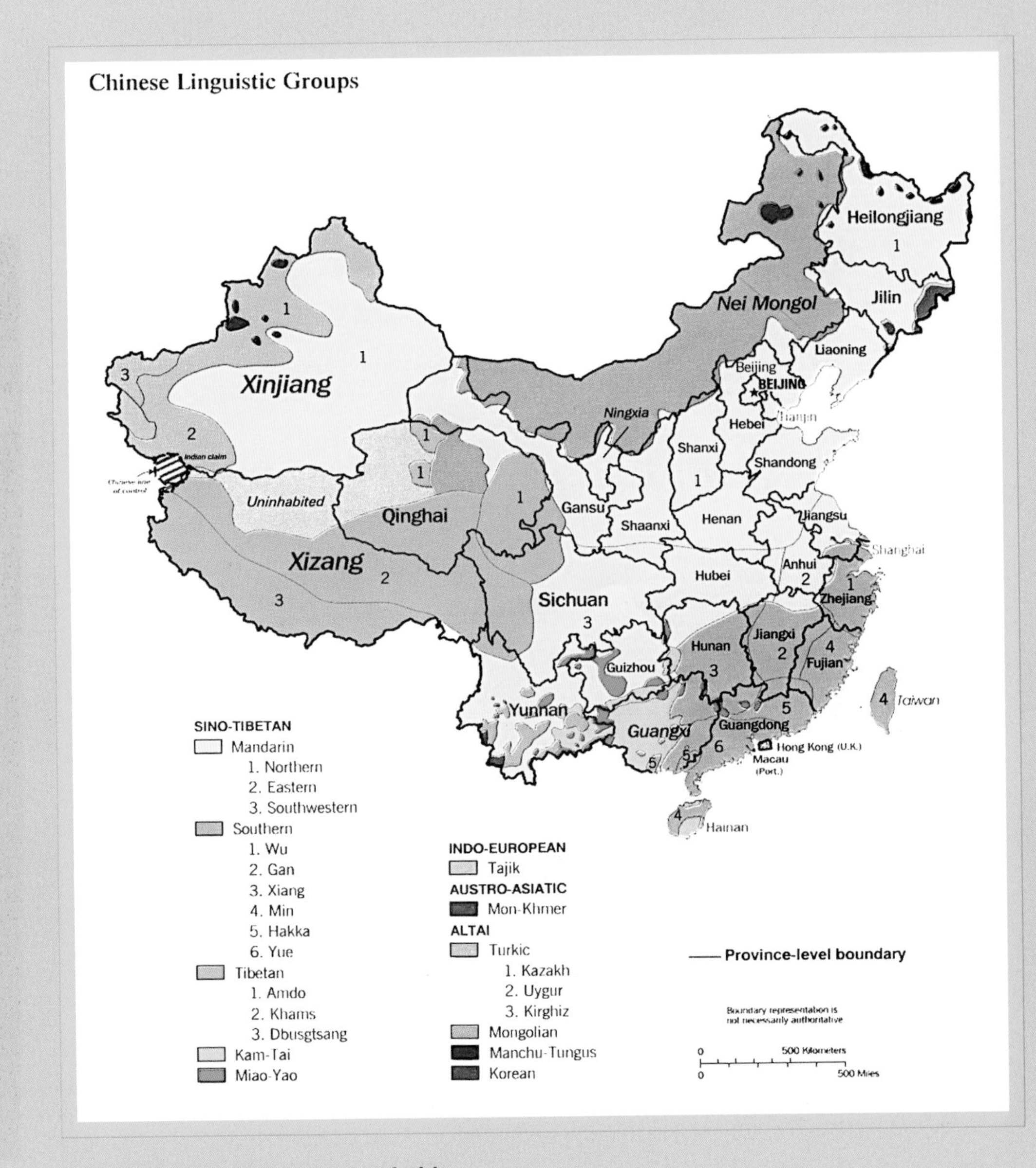

Map of the Linguistic Groups of China

A popular image of China is one of rolling, terraced rice paddies, which the people work tirelessly to feed the huge Chinese population.

1

Introduction

China has one of the world's oldest continuous civilizations. It is thought that some form of organized society began around 2000 B.C., and throughout the centuries, China has made significant contributions in philosophy, religion, science, math, politics, agriculture, writing, and the arts. With a population of 1.3 billion and land area of 3.7 million square miles (9.6 million square kilometers), China dominates not only Asia but the entire world. It has about 22 percent of the total world population, but only some 6.5 percent of the world's land area. The popular image of China is one of large numbers of people using tremendous energy and effort to work and modify the land. The latest massive effort is to realize the dream of the great early-twentieth-century Chinese nationalist, Dr. Sun Yat-sen. This is the idea of controlling the mighty Chang Jiang (Yangtze River). The present-day construction of the world's largest hydroelectric

dam at the Three Gorges is the fulfilment of this dream.

One key to knowing about China is to understand its diversity. This diversity is best recognized in the features of China's landscape. China is a rugged country, with mountains, hills, and plateaus occupying about 65 percent of the total land area. The highest peak in the world, Mount Everest, stands on the border between China and Nepal. Unfortunately for the Chinese people, only 10 percent of its land is available for cultivation, compared with 40 percent in the United States. Thus the people of China, with less than 7 percent of the world's arable land, constantly work to feed 22 percent of the world's population.

China occupies a strategic position in Asia. It is situated between the largest landmass, Eurasia, on the north and west, and the largest ocean, the Pacific, on the east. Mainland China extends from 20° north to 53° north latitude, covering 2,300 miles (3,680 kilometers). It stretches from 74° east to 134° east longitude. This longitudinal distance of 3,230 miles (5,170 kilometers) means that when it is noon in the east, it is only early morning in the far western part of the country. China's land boundaries total some 14,000 miles (22,000 kilometers), and China shares national boundaries with 14 different countries. It has an extensive coastline of 9,000 miles (14,500 kilometers), which includes the territorial waters of the Bohai Gulf and three neighboring seas—the Yellow Sea (Huang Hai), the East China Sea (Dong Hai), and the South China Sea (Nan Hai).

Perhaps no country in the world historically has interacted with its rivers more than China. The growth of civilization in the country has centered on three great river systems, all of which flow from west to east. The northern quarter is drained by the Huang He (Yellow River), which runs approximately 3,000 miles (4,700 kilometers) from the western territory of Tibet to its mouth in Shandong Province. The middle half of the country is drained by the Chang Jiang (Yangtze River), which also originates in Tibet. It is the longest river in China at 3,900 miles (6,300 kilometers), and has ten times the water discharge of the Huang He. In the near future, it will be easily navigable by ship from Shanghai in its delta, to the far western inland city of Chongqing. More than half of China's population lives in the Chang Jiang basin, because it includes the richer and more productive land in the country and receives adequate rainfall. The southern quarter of China is dominated by the Xi Jiang (West River). It is the shortest of the three major rivers, flowing approximately 1,650 miles (2,655 kilometers) before merging with the Zhu Jiang (Pearl River) in the delta. The two noted cities of Guangzhou (formerly Canton) and Hong Kong are situated at the mouth of this river.

Over time, these great river systems have been very important to China. They were the cheapest and most practical form of transportation, and were an important source for irrigation and energy. The river valleys also provided fertile alluvial soils for the surrounding level land. Alluvial soils are stream-deposited material. As a result of these factors, China's population is concentrated along these rivers and reaches its highest and most extensive density at their mouths.

Central to any knowledge of China is an understanding of the location and character of its people. Nine-tenths of the population are of the Han ethnic group, and the remaining one-tenth is divided among 55 other distinct ethnic groups. Traditionally, the Chinese family functioned as a unit and worked the land by plowing, seeding, thinning, and harvesting. It was the peasant-agriculturist who was the principal bearer of Chinese civilization, and thus Chinese history is tied to this continuously expanding agrarian society. More than 900 million Chinese peasants still work the land. However, one of the most notable changes in the last 20 years has been the increase in the number of cities in China, as people move to find work. Large cities help attract foreign investment, and the country plans to expand hundreds of towns into new cities. By 2025, the urban population is expected to comprise 55 percent of the total population. A further incentive to urban expansion is the relaxing of the residency permits, or *hukou*. Without these, one could not move to another location without approval from local government authorities.

China's population has long enjoyed strong traditions and values. Confucius, whose teachings and writings still influence Chinese thought, lived during the Zhou Dynasty about 2,500 years ago. The Confucian classics were the guide for Chinese civilization, highlighting education and family as the foundation of society. Taoists believed that people should renounce worldly ambitions and turn to nature and the Tao—the eternal force that permeates everything in nature. Buddhism became a very powerful belief system for the Chinese. It provided a refuge in the political chaos that followed the fall of the Han Dynasty (206 B.C.–A.D. 220). By the A.D. 400s, Buddhism was widely embraced throughout China. Over the centuries, openness to new ideas fostered the emergence of many Chinese inventions and discoveries.

The Chinese government has always been characterized by some form of central authority, dating as far back as the Xia Dynasty of 2200 B.C. Subsequent dynasties reinforced cultural unity and continuity for the Chinese civilization. One dynasty succeeded another through warfare, but with only occasional intrusion by forces outside of China. The lack of outside contacts allowed the Chinese to develop one culture across many regions with a strong sense of national identity. A series of emperors served as political leaders supported by well-equipped armies. The Chinese people believed their emperors ruled by "A Mandate of Heaven." These dynasties continued over the centuries until the opening of China to the West in the 1800s, with the eventual emergence of the Communist leadership and government in 1949.

China's foreign and economic policy took significant twists and turns in the twentieth century. This is apparent in the admission of China to the World Trade Organization (WTO) in December 2001 after some 15 years of negotiations. This will

Despite China's massive size, the people have maintained a culture that has continued for thousands of years. These people performed a traditional Chinese drum dance to celebrate the fiftieth anniversary of China's Communist government in 1999.

likely change the balance of power in Asia for the foreseeable future. China's entry into the WTO is generally regarded as the most significant event since the country adopted a policy of openness and reform in the late 1970s. Some equate it to China's being given a Security Council seat in the United Nations in 1971. Further, a China-Association of Southeast Asian Nations (ASEAN) free-trade area will create a regional trading bloc, similar to those that exist in Europe and the

China occupies a strategic position in Asia because it is situated between the largest landmass, Eurasia, on the north and west, and the largest ocean, the Pacific, on the east. It shares national boundaries with 14 different countries and has an extensive coastline of 9,000 miles (14,500 kilometers).

Americas. Thus, China's economic might, with the government being the most important player, will become very dominant both regionally and globally. China has become part of the global economy, and sharp increases have occurred in tourism, trade, and foreign investment. Exports comprise 80 percent of China's foreign currency exchange, and light industrial products such as small appliances, clothing, and footwear have overtaken agricultural products in importance. This growth of China's economic might translates into increased military strength. Always a concern in such a picture is the claim that mainland China makes for ownership of the offshore island of Taiwan. However, this may be countered as the government of Taiwan continues to ease restrictions on the money flow to the mainland. This should link the economies of the small island democracy and the world's most populous marketplace.

French Emperor Napoleon Bonaparte once remarked that China was a sleeping giant and that whomever awakened it would be sorry. Today, China is awake and is ready to take a leading role. The country with the world's largest population also has a wide range of natural resources, from fertile soils to forests to fuel and metals. Its human resources are enormous and its economic potential seems unlimited. It has a military of some 3 million soldiers, with an additional 1.2 million reserves. It has detonated nuclear bombs. The government has stated that space exploration will become "as essential as electricity." In 2003, China launched its first astronaut (taikonaut) and made plans to send an unmanned mission to the moon in 2007. The country is embracing limited capitalism within an authoritarian form of government.

It is important to learn as much about China as possible and to better understand and appreciate its character from every possible angle. It is a country whose history is so unique that it defies the Western mind. At times, the achievements of China have been obscured and forgotten, given the longevity of Chinese civilization. China will not disappear, but will most likely become one of the dominant world powers in the coming years. It is vital, therefore, in today's interrelated world, to know a great deal about this remarkable country.

China's landscape is extremely varied and beautiful. The Yangtze River is seen here flowing through Tiger Leap Gorge, which is the deepest in the world. Along with the Yellow River, the Yangtze has traditionally been a center of Chinese civilization.

The Natural Landscape

There are many ways of looking at China's landscape. However, it is important to remember that both the natural and human landscapes have been continually changing over the centuries. Well-known geographer G. B. Cressey once remarked that the most significant element in the Chinese landscape is not the soil, vegetation, or climate, but the people. In this very old land, one can scarcely find a place untouched by human activity. For example, the Huang He, or Yellow River, has flowed in several major channels during recorded history. It has been altered by floods and channelling projects. Large parts of China's south and north are now stripped of trees by human encroachment. Areas of Chinese culture and settlement have extended over a span of 80 centuries from the original core in the middle Huang He to the nation's present-day borders.

It is certain that the natural landscapes of China have greatly affected its historical development. Mountains comprise over one-third of China's area, and such rugged terrain has hindered cultural and ethnic blending for centuries. Besides the numerous high mountains, there are also extensive plateaus, rolling hills, inhospitable deserts, enclosed inland basins, and extensive fertile low plains. Such variety led the Chinese people to use the land in many different ways.

One fundamental way to understand the natural landscapes of China is to divide the country into two regions—China Proper (Inner China) and Frontier China (Outer China). The boundary marks a contrast between a settled, frequently irrigated, and intensely farmed area, and a marginal dry-farming area supplemented with agricultural enclaves, like oases of the northwestern deserts. The area is defined by a contrast between a section of huge cities and settled villages based on intensive agriculture, and an area of animals grazing on the plateaus. It is in Frontier China that groups such as the Mongols, Kazaks, Tibetans, Uygurs, Manchus, and other minority groups, live. In contrast, overwhelming numbers of Han Chinese live in China Proper. Just as sophisticated rice culture helped shape the agricultural scene of the southeast and east, so the care of sheep, goats, camels, horses, and cattle helped shape the nomadic lifestyles in the north and the west. These peoples were not tied to the land as were their eastern counterparts. The two basic regions are roughly comparable in area, but only 5 percent of the population lives in Frontier China. Two-thirds of the country is virtually uninhabited, and this makes the density of settled places even more pronounced.

China Proper

China Proper (eastern China), a region of hills and plains, comprises the cultural, agricultural, demographic, and industrial core of the country. It is a gentle land of alluvial

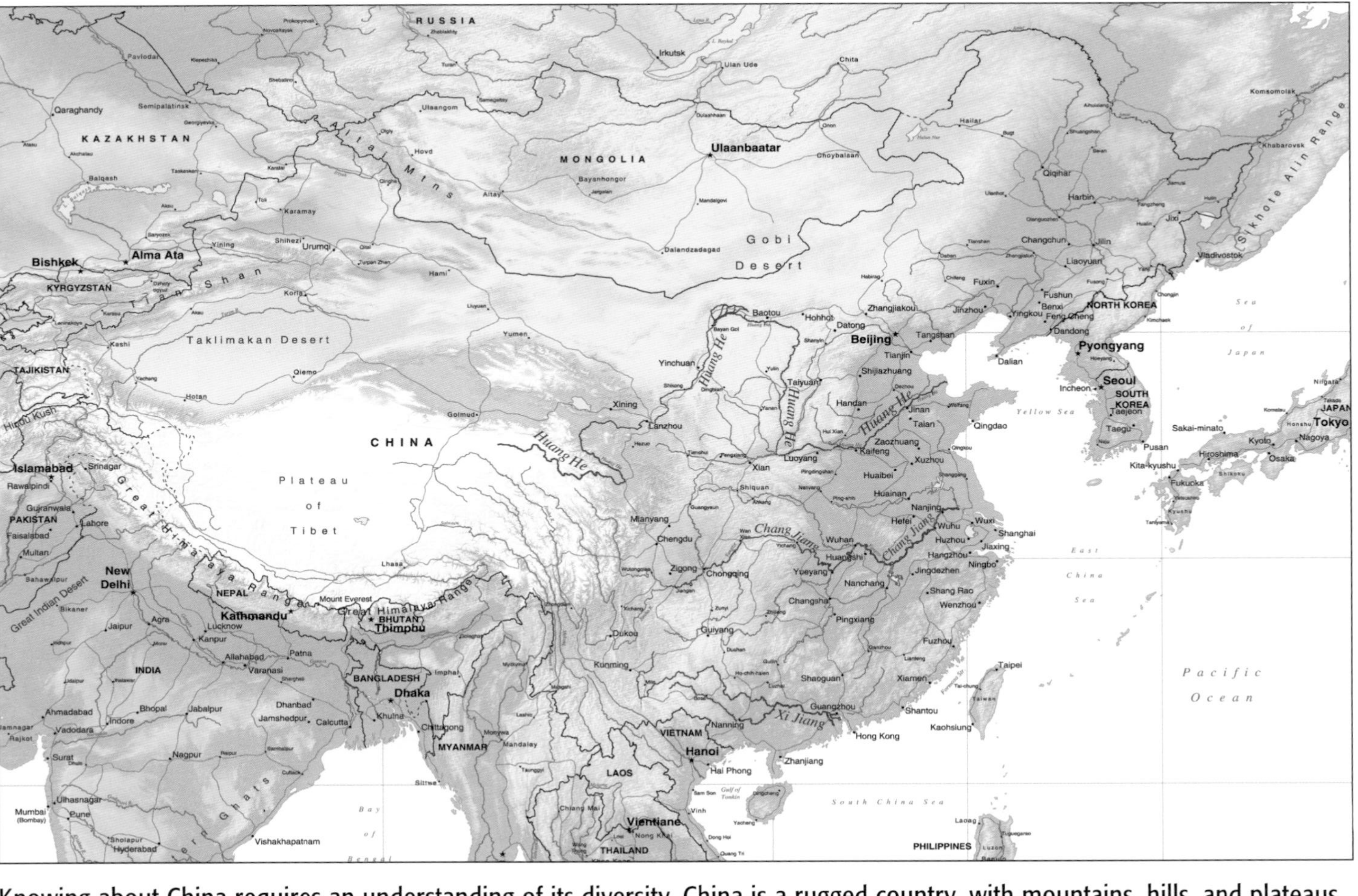

Knowing about China requires an understanding of its diversity. China is a rugged country, with mountains, hills, and plateaus occupying about 65 percent of the total land area. Perhaps no country in the world has interacted with its rivers more than China. The growth of civilization in the country has centered on three great river systems: the Huang He (Yellow River); the Chang Jiang (Yangtze River); and the Xi Jiang (West River).

plains, river valleys, and rolling hills. It is this part of the country that has been worn down by centuries of human occupancy and activity. The complex network of water channels in this area served to develop shipping routes that fostered coastal development. It contains four distinct landscape regions: the Northeast, the North-central, the South, and the Southwest.

Northeast Region

The Northeast region, referred to as *Dongbei*, is better known as Manchuria. This area was home to the Manchus and was later dominated by foreign powers, such as Russia and Japan, in the nineteenth and early twentieth centuries. It features the large Manchurian Plain, a rich and productive agricultural region that also holds important resources. Manchuria contains three of China's provinces, and includes the Xingan Mountains to the west and north, and the Chang Bai Shan to the southeast, along the border with North Korea. These are important timber-producing areas of oak and hardwood. Not fully developed until the late nineteenth century, the area now ranks first in terms of land quality and second in productive acreage for the entire country. It accounts for about 20 percent of China's farmland.

The climate of the Northeast Region allows only one crop to be produced per year. Spring wheat, corn, sorghum, millet, soybeans, and sugar beets are the chief crops. Low temperatures, thin soils, frost, spring drought, and summer-autumn floods are the main natural hazards facing this region.

The Amur River forms part of the northern Manchurian boundary with Russia. Its largest tributary, the Songhua River, along with lesser tributaries, form one of the great river networks in the country. It is second only to the Chang Jiang in annual discharge. As a result, there is no lack of irrigation when required, and the surrounding rugged landscape has potential for hydroelectric power. The Second Songhua River is famous for its Fengman Hydro facility, along with its huge reservoir, Songhua Lake. River boundaries are a common feature of this region. The Yalu and Tumen rivers separate this part of China from North Korea.

The lower Liao River marks the southern part of Northeast China. It empties into the Gulf of Bohai near the Liaodong Peninsula. This area has a more temperate climate and can sustain three crops over a two-year period. It is an important place for the production of fruit, fish, and a variety of silk. The area is experiencing dramatic environmental problems, such as soil erosion, caused by deforestation, and air and water pollution.

The Northeast has undergone numerous periods of promise and despair. Japanese occupation between 1931 and 1945 was ruthlessly exploitative. Later, from 1950 to 1970, the Communists made the development of resources in the Northeast a priority. In part, this growth was tied to the considerable mineral wealth. Iron ore deposits and coal fields are concentrated in the Liao Basin. Other plentiful metals include aluminium ore, lead, and zinc. It is also an area with extensive oil reserves. Plentiful resources make this area the second largest iron and steel center in China. Harbin is the main commercial center because of its strategic position, where five rail lines converge.

North-Central Region

North-central China is dominated by the North China Plain, which fronts the Bohai Gulf and the Yellow Sea. The physical landscape almost lacks elevation, but is dominated by two major river valleys. The Huang He (Yellow River) and the Chang Jiang (Yangtze River) to the south are the main lifelines of the region. This area has historically been the center of Chinese civilization. Today, it remains one of the world's most heavily populated agricultural areas. It has been the cultural, political, and economic center of China for thousands of years, and six ancient capitals of China are located in this region.

The Loess Plateau is important in this region. It is the cradle of Chinese agriculture. For 80 centuries, millions of peasants have lived on and cultivated this land. However, misuse, overuse, and overgrazing of the slopes have led to very destructive soil erosion, some of the worst in the world. Nowhere else in China has the struggle by the people to overcome nature been more apparent.

Loess is fine, wind-borne yellow-grained dust with a top layer that forms an easily cultivated soil. Millions of cave dwellings were excavated from loess because it is able to stand in vertical cliffs. These dwellings were warm in winter and cool in summer. Loess can be treacherous, however, quickly turning to impassable mud when flooded. In such a condition, it is called "liquid mud." The landscape can constantly change as loess is picked up as silt by rivers. The abundance of this yellow soil in the waters of the Huang He give it its name, the Yellow River. As the river descends from the Loess Plateau to the broad eastern plain, spring flooding often occurs. The river is also referred to as "China's Sorrow," because of the endless tragedies brought about by these floods over the centuries. Yet, it is the silt deposits brought by the flooding that has made this North China Plain such a rich agricultural area.

Diking and damming barriers have always been part of river systems in this part of China. Records show seven catastrophic shifts of the river channel, and 26 large-scale shifts along the Huang He from 600 B.C. to A.D. 1950. One breach in 1938 left 900,000 dead and 13 million people homeless. Major floods can force one million people to work at moving earth and sand bags to try to hold back the waters. Because of diking and the continual flow and build-up of silt, in some places the river is higher than the surrounding plain. This makes the flooding problem worse. Dam building on the Huang He also reduces the water flow, and incoming tidal water makes adjacent land too salty to farm.

There is an old Chinese saying that water is the lifeblood of agriculture, while transportation is the blood vessel of the economic structure. In this part of China, the people have used water and rivers skillfully over the centuries. Rivers were the original highways for transporting people and goods, and remain so today. Open sea transportation to Korea began as early as the Qin Dynasty (221–206 B.C.). Inland waterways have long been important. Canals were dug as early as 500 B.C. The famous Grand Canal, at a length of 1,550 miles (2,500 kilometers), was built from Beijing to Hangzhou. This is about the same distance as that between New York City and Denver, Colorado. The Grand Canal served as a major water and commercial link for the movement of goods and people between southern and northern China.

South Region

A significant geographic line in the South Region is formed by the Qin Ling Mountains, north of the Chang Jiang. These separate the northern dry wheat-growing areas from the southern warm and humid rice-growing areas. The northern slopes of the Qin Ling are short and steep, and the foothills are marked sharply by a great fault line. The southern slopes are rather long and gentle. Almost half of the land is original forest where the giant panda and the golden-haired monkey still survive.

South of the Qin Ling Mountains lies the Sichuan Basin, which contains the Chang Jiang. It is one of the largest inland basins in the country, occupying nearly 50 percent of the almost 360,000 square miles (933,000 square kilometers) of Sichuan Province. The red sandstone soil mixed with purple shale give the basin its common "Red Basin" name. Sichuan Province is one of the most populous provinces in China, with up to 100 million people. Surrounded by mountains,

Because the terrain of China was often difficult to travel, rivers and waterways were the easiest way to move goods and people around the country, even in ancient times. To make the waterways more convenient, the Chinese dug canals. The most famous of these is the Grand Canal, which runs north and south for 1,550 miles (2,500 kilometers).

the basin enjoys a milder winter than parts of the middle and lower Chang Jiang plains. Relative humidity can be high during the year, and parts of the basin can experience foggy days. The Sichuan Basin has been famous for centuries as an attractive and productive land. It has been intensively cultivated for over 2,000 years and is one of the most concentrated rice paddy areas in the world.

The nearby Chengdu Plain has long been the base for commercial grain and rape seed oil in Sichuan Province. The triple-cropping system (two crops of rice, followed by one crop of rape seed or wheat) predominates. Irrigation greatly supports agriculture, and the famous Du Jiang irrigation system built in 250 B.C. makes full use of the available rivers. This extensive irrigation system has been operating uninterrupted with only improvements and enlargements.

Within the basin are the magnificent Wushan Mountains and the spectacular Three Gorges. This is where the Chang Jiang, or "Long River," is forced to flow through a narrow 150-mile-long (240 kilometers), steep-walled valley no greater than 350 feet (107 meters) wide—slightly more than the length of a football field. The Sanxia, or Three Gorges Dam Project, is being built here to alleviate downstream flooding, produce inexpensive hydroelectricity, and improve upstream navigation.

As the Chang Jiang leaves the restricted confines of the Three Gorges, it meanders sluggishly across the flat terrain of its middle and lower course. Rivers and lakes become more widespread, forming a scenic and productive area known as the lake country. Numerous low mountains and hills encircle this middle plain. The lakes act as flood reservoirs for the Chang Jiang during the high-water summer monsoon season. (Monsoons are winds that are wet in summer and dry in winter.) These lakes, such as Dongting Lake, may well shrink when the Three Gorges Dam is completed. The lower plain of the river lies less than 10 feet (3 meters) above sea level, and this wetland environment has a patchwork of rice paddies and fish farms. These are linked by an extensive network of streams and canals. The force or speed of the water flow of the Chang Jiang is 17 times greater than that of the Huang He. The Chang Jiang is the third longest river in the world, about 200 miles (322 kilometers) shorter than the Nile, the longest. Chongqing, Wuhan, and Nanjing, cities located along this river, are often referred to as the "ovens of China." They experience very high summer humidity. The lower part of the Chang Jiang lacks the protection of nearby mountains, and thus can have colder winters, light snow, and periodic heavy fog.

This part of China is influenced by a humid, subtropical climate. When high temperatures and heavy rainfall come together, an opportunity exists for the intensive triple-cropping system of two crops of rice, followed by one crop of wheat or barley. Frost damage can be a problem owing to the east coast location.

The delta of the Chang Jiang is one of the oldest cultivated and irrigated areas in all of China. Competing for land is the city of Shanghai with an exploding population of 15 million and an expanding industrial complex. The area is an example of how urban development is causing rural land to disappear.

In the extreme southeast lies another productive area in the delta of the Pearl River and Xi Jiang (West River), near Hong Kong. This region has ample water and favorable high temperatures. Land that can be used for agriculture is limited by rugged terrain, however, when compared with North-central China. The river plains in the south can support triple cropping with rice, sugar cane, and sweet potatoes. Rice paddy fields and large fish farms also coexist here.

Hong Kong, a city of 7 million people, with a huge commercial and shipping center, and nearby Macao, with half a million people, compete for land in this delta region. Urban expansion and pollution are constant problems.

The surrounding low mountains have fared badly from deforestation and soil erosion.

Southwest Region

The Yunnan-Guizhou Plateau of the southwest is situated between southeast China and the Tibetan Plateau. It is China's most physically stunning region, dotted with high ridges and deep gorges. Elevations range from 3,000 to 7,000 feet (914–2134 meters). The region is quite wet because of the Indian and Pacific monsoons. The border areas with Southeast Asia are underdeveloped and have an extensive rainforest.

Part of the stunning landscape was created by water erosion on the thousands of square miles of limestone rock, or karst, near the city of Guilin. The nearby Li River wanders through haystack-shaped hills that stand 1,200 feet (366 meters) or more above the surrounding lowlands. These spectacular hills are found in very few places in the world. Intricate underground caves, some of which contain water channels, are another feature of this limestone area.

Frontier China

Frontier China has gorgeous but quite inhospitable scenery. It includes the world's highest mountain, Mount Everest, which stands 29,035 feet (8,850 meters) above sea level, and the world's second lowest place below sea level on dry land, the Turfan Depression, which is 505 feet (154 meters) below sea level. Frontier China also contains the world's most barren deserts, the Gobi and Taklamakan. Huge swamps such as those of the Qaidam Basin are found in the northern part of the Tibetan Plateau, with virgin forests and endless grass-covered steppes scattered throughout. Much of Frontier China is an inland drainage area in which rivers both begin and end their journeys. For this reason, Frontier China was cut off from the extensive system of inland waterways that developed throughout eastern China.

Frontier China and China Proper were effectively joined together politically under the Manchu (or Qing) Dynasty, mostly during the eighteenth century. This process created what is now recognized as the nation of China. During the last two centuries, the Han Chinese migrants extended their influence into Frontier China, and converted much of Manchuria from forest and pasture into farmland.

The one great route that joined China Proper to Frontier China was the famous Chinese trade route, the Silk Road. It began around 138 B.C. Later, it was extended, and eventually linked the Chinese and Roman empires. Its use only declined in the sixteenth century. Chinese silks and products such as jade, pottery, fruits, and paintings, as well as ideas, were sent to the Roman Empire in exchange for gold, glassware, wool, and linen fabrics. The Silk Road looped south and north of the scorching Taklamakan Desert and rose high through the mountain passes across the Pamirs.

The Himalayas in the southwestern part of the country effectively seal off China from India, Nepal, and Bhutan. In the shadow of these mountains to the north lies the massive Tibetan Plateau. Slightly smaller in area than the 48 contiguous United States, the 2.5- million-square-mile plateau (6.5 million square kilometers) has an average height of 14,765 feet (4,500 meters) above sea level. It accounts for one-fourth of the Chinese landmass but contains only one percent of China's population. Also included in this region are other great mountain ranges: the Karakorum *Shan* (*shan* means "mountains"), separating China from India and Pakistan; and Kunlun Shan, separating the plateau from Xinjiang Autonomous Region to the north.

The Tibetan Plateau is where some of the great rivers of Asia have their source. The Huang He (Yellow) and Chang Jiang (Yangtze) flow into China Proper. The Mekong, Irrawaddy, Brahmaputra, and Salween rivers also have their source on the plateau.

North of the Kunlun Shan lies the oval-shaped Qaidam Basin. It is the transitional area between the frigid Tibetan Plateau and the arid northwest. A series of land types are found within the basin, including a desert landform feature known as a *yardang*. The yardang is made up of elongated ridges of rock formed parallel to the prevailing wind direction and separated from a similar formation nearby by a shallow furrow. This is the most extensive yardang area in China. It is a very rich salt-mining area, and nearly all of China's potassium reserves are found here. The basin also has extensive petroleum reserves.

To the far northwest of the Tibetan Plateau are the Xinjiang-Mongolian Uplands. This is another environmentally inhospitable region. Two great mountain ranges, the Tien Shan and the Altai Mountains, intrude into this region and encompass several large depressions or basins—the Tarim Basin, the Turfan Depression, and the Jungar Basin.

The Tarim Basin contains the only warm temperate desert in China. Several crops may be grown where water is available. The basin runs more than 1,000 miles (1,609 kilometers) west-east and about 400 miles (644 kilometers) north-south. Within this basin lies the barren Taklamakan Desert, which measures 127,414 square miles (330,000 square kilometers) in area. It is the largest sand desert in China and the second largest in the world. The name of this so-called sea of death translates as "go in, not come out." It is thought that this area has great potential for oil deposits and there is evidence of natural gas and minerals. On a huge salty marshland, just east of the Taklamakan, lies Lop Nor, the site where China's 32 underground nuclear bomb tests were conducted between 1964 and 1988.

The Turfan Depression was the site of an ancient underground irrigation system, called a *karez*, developed between the eighth and tenth centuries. This system was devised by the Uighur people, who later converted to the Muslim faith. The karez transformed the area into a garden of grapes, melons, and palm trees.

Two important gateways from the Jungar Basin pass through mountain ranges to Kazakhstan in the west. The Jungar Basin is noted for the seasonal movement of people and animals to fresh pastures, in some cases according to rainfall fluctuations.

Xinjiang Autonomous Region has large open basins surrounded by mountains, and can become very hot in the summer. It is so far inland that there is hardly any rainfall, and settlement is confined to the foothills and oases that surround the depressions. At 617,760 square miles (1.6 million square kilometers), it is the largest province in China. Twenty million people live there.

North of the Xinjiang-Mongolian Uplands lies the Altai Mountain System. It stretches northwest to southeast more than 1,240 miles (2,000 kilometers) along the Sino-Russian-Mongolian border. Only the middle section lies within Chinese territory. The southeastern edge eventually merges into the vast level expanse known as the Gobi Desert. These mountains are the source for rivers that flow into the Jungar Basin.

Part of the Gobi Desert lies in the independent nation of Mongolia, but most of it extends into Inner Mongolia. It is the world's highest desert in both elevation and latitude. As a result, it is one of the coldest of all deserts. It is thought that the Gobi provided the source material for the vast stretches of yellow loess that cover much of northern China. This area, with its interior continental location, is isolated from strong winds and moisture-laden storms. However, there is enough wind to bring some moisture to support low shrubs and grasses, and to promote extensive raising of livestock.

To the south of the Gobi Desert lie the Ordos Plateau and Desert. The region is easily defined. The Great Wall is to the south and the other three sides form the loop of the Huang He as it flows north, east, and then south. The middle Huang He River encircles the region in a great arc. Access to water allows irrigation and fertile soils. This is one of China's most threatened areas, vulnerable to environmental change and desertification—the spread of desert conditions and desert lands into semiarid, or short grasslands, or even woodland areas, caused by reduced and uncertain rainfall and expanding human settlements. This has become a major problem for China, as the desert has migrated eastward in parts to within 50 miles (80 kilometers) of Beijing, the nation's capital. Occasionally, sandstorms can even close down the Beijing airport. The southern border of the desert has now also encroached upon the northern margin of the fertile Loess Plateau. Since 1949, it is estimated that one-fifth of China's farmland has been lost to soil erosion or economic development.

Climate

Five major climatic zones have been identified in China. These range from cold temperate to subtropical and tropical. The country also experiences climatic extremes. Over 30 percent of China is almost completely arid, but in other parts of the country, there are tropical rain forests. The weather patterns are highly seasonal, largely influenced by the annual monsoon cycle that impacts the southeastern coast.

The mountain topography modifies the weather systems to produce very different regional conditions. The biggest problem in the north and west is the lack of rainfall. Large areas receive less than 5 inches (12.5 centimeters) per year and agriculture, if possible at all, relies heavily on irrigation. East China receives enough rainfall for farming, but the Qinling Mountain range marks a geographic line. South of the range, there is a surplus of water and double cropping is evident. North of this range, there is insufficient water and heavy reliance on irrigation from rivers. Monsoon rains can be unreliable, so rainfall varies from year to year. Droughts and floods are always a part of China's climatic picture and a challenge for its people. The management of water resources has been a major concern for farmer and government alike.

Typhoons can be a source of moisture from July through October. They develop in the western Pacific and strike the southern and eastern coasts, impacting China more than any other country in the world. There are about seven typhoons per year. They can move inland some 300 miles (483 kilometers) and last several days, releasing great quantities of water. Typhoons bring winds that can reach 200 miles (322 kilometers) per hour, causing severe damage and flooding. Most weaken within several hours after touching land, however. The interior part of the country receives winds that are dry, having lost all moisture on the long journey from the coast.

Regional temperature varies more in winter than in summer. The northeast and west regions and interior locations have low average winter temperatures, in the -22°F (-30°C) range. North of the Qinling Mountains, the winter temperatures are below freezing. Beijing can average 23°F (-5°C) in January, while Guangzhou near Hong Kong in the south averages 59°F (15°C). Because of the mountain elevation and inland location, winters in western parts of China are below average in temperature for their latitude. Summers are excessively hot and humid, except on the Tibetan Plateau.

Earthquakes and Volcanoes

China has kept records of destructive earthquakes almost as long as history has been written. Most of these earthquakes are caused by the many geologic fault lines found throughout the country. The earth's surface is composed of 12 giant rock plates that are 70 miles (113 kilometers) thick. These crustal plates are in motion and push against each other, triggering earthquakes. China sits on the Eurasian major tectonic plate, which borders three other major tectonic plates in the region. The northward-moving Indian-Australian plate collides with China in the Himalayan region, pushing the mountains upward. Just east of China, the Eurasian plate intersects two other major plates near Japan. This is part of the Pacific Ring of Fire, a circle of earthquake and volcanic zones that surrounds the Pacific Ocean.

The Chinese have always been interested in earthquakes, and in A.D. 132, they devised the first seismograph to measure and record vibrations in the earth. Shaanxi Province, in 1556, experienced a devastating earthquake (believed to be higher than 8 on the Richter scale), that killed more than one million people. This earthquake was the worst natural disaster known to humankind. Similar large-magnitude earthquakes have struck the region over the years. Chinese scientists have compiled a 3,000-year-catalog of earthquakes, and 500 destructive ones have occurred during the last 1,000 years. One of every 16 earthquakes that hit China reaches a magnitude 8 or greater on the Richter scale. Since 1900, 48 deadly earthquakes have killed a total of more than one million people in China. In 1976, Tangshan, an industrial city 85 miles (137 kilometers) east of Beijing, lost up to 750,000 people in a reported 8.2 magnitude earthquake. The lack of earthquake-proof buildings throughout the country has made the Chinese people very vulnerable when earthquakes strike.

China also has a number of volcanoes. The China–North Korea central mountainous border area saw a very large eruption in 1060, one of the largest worldwide in the last 10,000 years. It was the largest known eruption in history on the Asian mainland. Another occurred in 1702. Eastern China has also recorded other eruptions. As with earthquakes, the coastal area seems to be more active.

Problems for the Land

Centuries of agriculture have modified China's plant and animal life dramatically. It is thought that much of China was once heavily forested. Deciduous trees in the south gave way to coniferous trees in the north. Pockets of tropical rain forest are still found along the south from the interior

Yunnan-Assam border to offshore Hainan Island. Generally, grassland, steppe, and desert dominate the north and northwest, and the woodlands are confined to the central and southeastern parts of the country. The Tibetan Plateau has the typical extensive meadows and alpine vegetation common in high latitude locations.

The extent of deforestation over the last 50 years has had massive consequences for the country. A United Nations (UN) report calculated that 36.3 percent of China's healthy forests that are located near high-density populations are under intense pressure. Another 55 percent of China's forest area is threatened as economic growth results in an increasing demand for wood. The country is the second largest importer of timber in the world because of desertification that is particularly noticeable in the north and west. The government tried to bring 17 percent of the country back to forest cover by year 2000. Tree planting has focused on the upper reaches of major river systems such as the Chang Jiang, Huang He, and Liao to stabilize the land and provide needed timber. A large project involves planting trees along a protective green belt stretching 4,350 miles (7,000 kilometers) from the northeast to Xinjian province in the far west. This is known as China's Great Green Wall.

Wildlife

Even with intensive human occupancy, in some areas there is still enough wilderness to support varied wildlife. There are reptiles, elephants, tigers, monkeys, and the famous giant panda. Some notable surviving Chinese wildlife include the great paddlefish of the Yangtze, the small species of alligator in east-central China, and the giant salamander in western China. Diversity of animal life is greatest in the ranges and valleys of the Tibetan border. The giant panda is found most often near the Sichuan Province–Tibet border. However, the future for this animal remains bleak, because it has a particular habitat

China is home to an amazing variety of wildlife. Unfortunately, as a result of construction, pollution, and other problems, several Chinese species are endangered. Among these threatened animals is the giant panda, which requires a very special habitat and diet to survive.

and must consume about 45 pounds (20 kilograms) of a specific type of bamboo each day.

The country supports 15 percent of the world's mammal and bird species, and a remarkable number exist only in China. These include 400 types of fish, such as the Yangtze River sturgeon; 100 kinds of birds, such as the red-crowned crane; and 70 varieties of mammals, such as the golden-haired monkey.

The World Resources Institute identified some frightening declines in China's biodiversity. Some 30,000 plants; 1,100 birds; 394 land mammals; 340 reptiles; and 263 amphibians are known to be threatened. Only 7.7 percent of China's land area is set aside as nature reserves.

Locusts are perhaps China's most destructive species. The country has been plagued for centuries by more than 60 types of locusts that have caused major agricultural damage. The East Asian flying locust is a particular menace. The North China Plain has been hardest hit over the centuries, as locust invasions have been linked to droughts and floods.

Rats can also inflict severe damage on crops as well as humans. Nearly 200 species have been identified in China. The eastern monsoon farming area and the northwest arid zone, with extensive pastoral grasslands, are especially vulnerable to rats. Estimates show the rat population of China at around 15 billion, compared with 8 billion in India.

Coastal Features

The Chinese coastline—over 9,000 miles (14,500 kilometers)—has made the country both a continental and maritime nation. The China shore is bordered by some 5,000 rocky islands, the largest being Hainan and Taiwan. These islands have created a zone of inland seas that provided sheltered routes for early trade between China and the islands of the East Indies. Along the coast, China has felt the greatest political pressures exerted by Western Europe in the form of "treaty ports" during the nineteenth century. These were privileged enclaves, particularly Shanghai, that served foreign trade and investment.

China's overseas relations with the world were much less important historically than contacts made along its land frontiers. However, near the end of the first millennium A.D., there were improvements in Chinese shipbuilding. A powerful navy developed briefly during the Ming Dynasty, but inland defense was the priority.

Generally, the coast borders a broad continental shelf with water depths from 98 to more than 328 feet (30 meters to more than 100 meters). The central sea basin of the South China Sea is deeper in places. The extensive area between the Yellow Sea and the East China Sea contains a very broad and rich petroleum-bearing area that is 300 nautical miles (556 kilometers) wide.

The Chang Jiang and Pearl River deltas were developed for irrigation and fish farming. Besides the intricate fish farming, offshore and deep-sea fishing yield about 15 million tons a year, a very dramatic increase since the 1950s. The Chinese fishery is one of the largest in the world, accounting for 15 percent of the total global catch. The variety of fish and shellfish is limitless. Lobster, crab, mackerel, herring, eel, shark, sardine, and sturgeon are just some of the choices. Still, fisheries are a weak part of China's economy.

China has established 15 major seaports. Since the late 1970s, the country has given special attention to the coastal cities, establishing "special economic zones." These are manufacturing areas that use tax, tariff, and investment benefits to attract foreign investors.

This is only an overview of China's natural landscapes. Some argue that the landscape is of fundamental importance, not only to the past but also for the future. China's future actions will be governed, in part, by its huge land area. It is vital to know the character and composition of this land area to better understand and appreciate how and why the Chinese people interact with their land.

China has a unique culture that dates from the ancient imperial dynasties. These are some of the famous Terracotta Warriors—a group of 8,000 life-size statues built to guard the tomb of the emperor of the Qin Dynasty (221–206 B.C.).

3

China through the Centuries

Chinese civilization is ancient. The country had dynastic rule for most of its recorded history. This was a succession of emperors or rulers in the same line of descent governing the country. These rulers established the basic pattern of imperial bureaucratic government that lasted for centuries until the twentieth century. Usually, dynasties changed through conflicts and battles, and capital cities often changed as well. Among the major dynasties were the Han (206 B.C.–A.D. 220), Tang (618–907 A.D.), Song (960–1279 A.D.), Mongol or Yuan (1279–1368 A.D.), Ming (1368–1644 A.D.), and Qing (1644–1911 A.D.).

China has the longest continuous history of any civilization in the world, but the exact origins of the Chinese people and their culture are unknown. Early humans are known to have lived in China half a million years ago. In 1929, the discovery of Peking Man in a cave southwest of Beijing on the North China Plain shows that people lived there around 500,000–210,000 B.C. Archaeological finds elsewhere in China suggest several similar cultures at the same time. It is likely that the Huang He Valley, like the river valleys of Egypt and Mesopotamia, supported settlement from very early times. It appears that several different ethnic groups and centers of early culture gradually mixed to produce the civilization that has continued over the centuries.

Archaeologists and historians argue that the first organized Chinese state did not appear until several dynasties in ancient Egypt and the city-states of Mesopotamia had come and gone. Once it developed, however, Chinese civilization proved quite long-lasting.

Discoveries have revealed the existence of a Neolithic culture at numerous sites in China. The Neolithic period dates from the latter part of the fourth millennium B.C. until the onset of the Bronze Age (2000–1400 B.C.). It was characterized by finely fashioned flint and stone, with the beginning of animal domestication, crop cultivation, and pottery making. The beginnings of agriculture, dated at 7000 B.C., developed in several areas in China. Early farming centers may include the Middle Huang He (Yellow River) and its western tributary, the Wei valleys, and the lower Chang Jiang (Yangtze River) Valley, and adjacent coastal plains of the southeast.

Early Settlement

Chinese legends mention gods and demigods in what are now Gansu and Shensi provinces of North China. From the legendary first man, Pan Gu, came a series of rulers, and finally, the emergence of the Yellow Emperor around 2550 B.C. This emperor is said to have expanded the boundaries of the empire. From 2300 to 2140 B.C., the famous rulers Yao, Shun, and Yu are mentioned. These are the first monarchs identified in the Shu Ching classics and were later regarded as model rulers by Confucius. Dates and details are uncertain, but Yu the Great is said to have solved flood problems and to have founded the first dynasty, the Xia (or Hsia), which possibly dated 2205–1765 B.C. Even its existence remains in question, but many Chinese scholars accept this as a historical reign. It is at this time that the system of hereditary kingship was created. This evolved into the centralized imperial system that lasted for several thousand years.

About 2100 B.C., the Chinese Bronze Age began and brought significant development. This was the age when bronze, made of copper and tin, was used to create objects. Later alloys were added, and then iron, a much harder metal, replaced bronze. At that time, Tang the Accomplished overthrew a tyrant king and founded another dynasty, the Shang or Yin Dynasty that lasted from 1523 B.C. to 1027 B.C.

There appears to be evidence to link parts of southeast China—the present-day provinces of Guangxi, Guangdong, and Fujian—to settlement areas farther south in the vicinity of Vietnam. Influences from Southeast Asia came to China during this Neolithic period and were important for many centuries.

Early rice cultivation formed the basis for settlement along the Chang Jiang and Huai River valleys. Excavations show that this occurred about 10,000 years ago. Rice cultivation spread down the river valleys and along the East China coast. Domestication of dogs, pigs, and water buffalo became an integral part of settlements. Pottery artifacts also pointed to a progressively refined culture, as did evidence of a writing system and wheeled chariots.

At this time, the idea of the Chinese state was forming. This period is known as the Shang Dynasty (1600–1050 B.C.). Exact dates are unknown, but the existence of this dynasty has been confirmed. The Shang kings controlled most of the Huang He plain, including Shandong, Hebei, and Henan provinces, and parts of the provinces of Shaanxi and Shanxi to the immediate west. Archaeological evidence indicates that the Shang state went well beyond these borders. The Shang period was marked by the sophisticated use of bronze and pottery molds, a culture that was quite different from other civilizations at this time.

Feudal States

During 1100–256 B.C., a rival tribal group emerged to challenge the Shang Dynasty. The Zhou people, located in present-day Shaanxi province, extended their power to the Huang He Valley and overthrew the Shang by force. A feudal system was put in place and became highly organized. The territory was divided into small states, governed by Zhou clansmen and supporters.

Zhou power lasted more than 800 years, longer than any other Chinese dynasty. Its decline came at the same time that people and animals migrated on the steppes in northern China. Around 900 B.C., armies were more mobile, with armed horse riders. Feudal states evolved to have independent armies competing with each other in incessant wars. The Zhou Dynasty, though depleted of its power, existed for another 500 years, from 771–221 B.C. Historical records mark these years as the Spring and Autumn period and the latter part as the Warring States period. It was a period of cultural advances in philosophy, the arts, and technology and a prelude to the first great imperial age in China.

The principal states of northern China were collectively known as Zhongguo, the "Middle Kingdoms." The Chinese considered their culture the center of the universe, and they felt they were surrounded by barbaric peoples outside their territory and isolated from other sophisticated cultures of India and western Eurasia.

This is the period of the development of great social and political philosophical thought for China. Confucius, who lived from 551 to 479 B.C., introduced systematic philosophy and ethics to the problems of government and the people. Mencius, his disciple, stressed the ideas of moral principles. Other philosophers with many different ideas emerged and influenced existing rulers. Taoism, another school of philosophy, also appeared. It was a more contemplative approach, compared with the rigid principles of Confucianism.

Confucius, the great Chinese philosopher, lived from 551 to 479 B.C. His teachings on ethics and moral principles influenced the Chinese Empire for many years.

First Emperor of the Qin Dynasty

By 221 B.C., King Zheng of Qin completed the conquest of all the Chinese states. He declared himself the first emperor of the Qin Dynasty. As the dynasty evolved, it expanded and consolidated its territory by following the guidance of ministers versed in legal philosophy. This helped the dynasty avoid direct battle. An area equal to that of China Proper was unified under one ruler.

The Qin Dynasty brought together a diverse cultural and linguistic territory. The 15-year Qin Dynasty was successful in centralizing authority and standardizing currency, transportation routes, weights, measures, and written script. The main problems for the empire were the threats from outside the northern frontier. The Chinese often protected towns and cities with large walls. Longer walls were also built, extending along the northern edge of the Chinese cultural center located to the south. Different construction materials were used based on local conditions—compressed earth or stone and brick. Eventually, the first Great Wall was completed in 214 B.C., through forced labor and with great loss of life. Today, the Great Wall stretches 4,500 miles (7,250 kilometers), the distance from New York to Las Vegas and back.

The burial ground for the Qin emperor is quite elaborate and unique. Discovered in 1974 by local farmers, subsequent archaeological diggings revealed a mausoleum of the now-famous Terracotta Warriors. These 8,000 life-size statues of soldiers stand in military formation. No two of the soldiers' faces are the same. Some people refer to these figures as the Eighth Wonder of the World.

Han Dynasty

The Qin Dynasty fell quickly. It marked the first time in Chinese history that a dynasty was toppled by a peasant uprising. A military officer who was a commoner declared himself emperor and created the Han Dynasty, which lasted 400 years from 206 B.C. to A.D. 220. Han China rivaled the Roman Empire in achievements and power. Technological advances included gunpowder, paper, porcelain, and the wheelbarrow. The teachings of Confucius became the cornerstone of state thought, and the people accepted the absolute power of the emperor. It was a period for advances in cultural and scientific achievements and a time to create a unified Chinese identity. Over the Silk Road trade route, Buddhism, the most important religion from outside China, came to the country from India around A.D. 65. Buddhism brought new ideas to China, including nirvana (heavenly bliss) and reincarnation. By A.D. 700, Buddhism, Taoism, and Confucianism coexisted.

The Han period saw great territorial expansion. The empire reached to present-day Xinjiang Autonomous Region in the far west, Korea in the east, and as far south as Vietnam. This was the first time that the Chinese empire bore some relation to its modern-day state and territorial limits. The population reached about 60 million.

The river basins of central and eastern China, with their large tracts of arable land, have always been the economic base for the Chinese state. This area sustained a large farming population. However, the north and west border lands were only controlled with great difficulty. Even today, Chinese rule in Tibet and Xinjiang Autonomous Region is continually contested by the Tibetans and Uighur Moslem peoples, respectively.

Fifteen emperors ruled during the Han period. As in earlier imperial dynasties, change came through uprisings and revolts. A number of kingdoms and dynasties arose, held and fought for power, and collapsed. This period of disunity lasted for 400 years from 220 to 589.

Sui Dynasty

The Sui Dynasty, 590–618, attempted to consolidate China and rebuild parts of the Great Wall. The Sui emperor also simplified and strengthened the bureaucracy, adopted a new legal code, and created a palace city near Xian. The dynasty's great construction feat was building the Grand Canal. This was an important south-north link between the rice bowl of the Chang Jiang and the capital at Xian. From Hangzhou north to Beijing, it linked several major waterways.

Tang Dynasty

The seventh century marked the start of the medieval period in Chinese history. Chinese culture reached a very refined and cosmopolitan level—literature, art, music and agriculture flourished. Under the Tang Dynasty (618–907), China expanded. The Turkish empires were defeated and the Tibetans became reliable allies. Mercantile cities, such as Guangzhou in the Pearl River delta near Hong Kong, were influenced by Islam. During the Tang period, the Uighurs, a Turkic-speaking people, created their own empire in central Asia, northern Xinjiang, and parts of Mongolia. They decided to ally with the Tang government rather than confront it.

The Tang dynasty witnessed the only woman in Chinese history to become a reigning empress. Empress Wu was a dominant power for 50 years (684–704). She was a great supporter of Buddhism and commissioned the famous Longmen Buddhist cliff carvings outside Luoyang, 185 miles (300 kilometers) east of the ancient capital of Xian. The 2,000 caves are shallow openings in the limestone cliffs where 100,000 Buddhist statues have been carved, including a Buddha that is 56 feet (17 meters) high.

The next 200 years, 704–907, saw the country split into regional political and military alliances. From 907 to 960, five short-lived dynasties succeeded one another. During this time, the defenses against the north were weakened and economic dependency on the south increased.

Song Dynasty

Finally, in 960, the Northern and Southern Song dynasties emerged for a 300-year period. The northern part of the country was well connected to the Grand Canal so that it could bring grain from the south. The country became very economically advanced, and industry and technology blossomed. Overseas commerce added greatly to the government revenues. Gunpowder, the magnetic compass, and fine porcelain, and movable type printing were made and used. Preoccupation with the arts and science probably caused a military decline. This eventually made the country vulnerable to the thirteenth-century Mongol invasion.

Mongol (Yuan) Dynasty

The period from 1279 to 1842 saw significant parts of China fall under foreign influence and at times foreign domination for the first time in the centuries. The Mongols from the north breached the Great Wall, and began their conquest early in the thirteenth century. By 1279, the Mongol (Yuan) Dynasty had been in place for about 100 years.

The rule of the famous Khublai Khan (1215–1294) set the tone for this dynasty. Beijing was selected as the capital, and influence spread to Korea, and as far west as Tibet and Burma. Between 1271 and 1292, explorer Marco Polo made his journeys to China. For 17 years, he resided in Beijing, opening the way to foreign influence.

Ming Dynasty

From 1356 to 1382, Mongol rule weakened and the Ming Dynasty was founded in 1368. It lasted 300 years. Between 1582 and 1610, the Jesuits brought Christianity to China. In 1514, Portuguese vessels sailed to the Pearl River delta, and in 1557, the Macau enclave formed. This colonial outpost lasted about 450 years, until December 1999.

The early Ming emperors were forward-looking and worked hard to rule and hold their power. They rebuilt the Great Wall, re-routed the Grand Canal to end near Beijing, and built a new southern capital at Nanjing. They established a powerful early navy and sent out maritime expeditions of diplomacy and exploration. Between 1405 to 1433, Chinese explorers went to the South China Sea, to Ceylon and India, to the Persian Gulf and to coastal East Africa, including Kenya and the offshore island of Zanzibar. Research suggests China's most famous navigator, Admiral Zheng He (1371–1435) completed the first circumnavigation of the world, beating Portuguese explorer Ferdinand Magellan by a century. This Chinese explorer, a Muslim eunuch, commanded a fleet of 300 ships and some 30,000 sailors. He is believed to have discovered America 72 years before Christopher Columbus. During these voyages, it is thought that he mapped the world. Despite these successes, Chinese expeditions were cut short in 1433. A conservative bureaucracy considered them a waste of money and resources. The Ming Dynasty withdrew from sea voyages and even banned coastal shipping for more than 150 years. China missed its chance to rival the Portuguese as a maritime power in the Indian Ocean by only 30 years.

Manchu (Qing) Dynasty

After the 300-year Ming rule, the Manchus from Manchuria gradually conquered China. They established the Qing Dynasty (1644–1911). This was the last of the dynasties, and brought China into the twentieth century, after 270 years of rule. The Manchus added Inner and Outer Mongolia, Tibet, Turkestan, and Taiwan to the empire's territories. During the latter part of Manchu rule, the British East India Company expanded. In 1793, Great Britain tried to establish trade relations with China. A secret market was established in China for western goods, especially the drug opium. The Opium War (1839–1842) brought China into conflict with British troops. China wanted its own control over trade with the West. The Treaty of Nanking was signed after China's defeat. China paid a heavy price for losing this war. Several enclave treaty ports were forced to open to British residence, and China lost Hong Kong to Great Britain.

The Taiping Rebellion of 1853–1864 was caused by anti-Manchu forces protesting economic hardships. The government troops, along with paid foreign troops, defeated the rebels. Twenty million Chinese died between 1850 and 1870 in prolonged fighting and economic disorder.

The mother of Emperor Tsai Chun, known unofficially as Empress Dowager Cixi, held the real power and ruled from behind the scenes from 1861 to 1908. The Qing Dynasty declined quickly. Under pressure from European powers and a newly risen Japan, the Chinese Empire was shrinking fast. Chinese dependencies such as Vietnam (Annan) and Burma were lost, as were parts of Chinese territory in eastern Siberia, parts of Manchuria, and the island of Taiwan. The principal beneficiaries were France, Great Britain, Russia, and Japan.

In the 1890s, China was in chaos. Foreign powers competed with each other to gain concessions from China. Economic hardship led the common people to protest in uprisings. The most serious was the anti-foreign Boxer Uprising of 1900, which caused a disastrous war and the occupation of the Chinese capital of Peking by the joint military force of eight foreign nations. Peace was restored only after China was forced to sign a humiliating treaty. The Qing Dynasty lingered on for another decade. It was toppled by a revolution led by Dr. Sun Yat-sen in 1912.

The Republic

A Chinese republic was founded in 1912, but the process of decline and political disintegration continued. Mongolia and Tibet declared their independence from China. The rest of the country soon plunged into incessant civil wars fought among military strongmen. The next 16 years, 1912–1928, are known as "the warlord period" in Chinese history. The central government was often ignored by regional warlords, while foreign powers took even greater control away from China. Another wave of revolution swept through the country, led by the

Nationalist Party founded by Sun Yat-sen in temporary alliance with the Chinese Communist Party. In 1928, the Nationalists defeated the warlords and set up a new central government under Jiang Jieshi (Chiang Kai-shek), the successor to Sun. This Nationalist government, with its capital at Nanjing (Nanking), ruled China until 1949, interrupted by a bitter war against Japanese invasions that lasted from 1937 to 1945.

Communist Mao Era

The long war against Japan sapped the energy and exhausted the resources of the Nationalist government. On the other hand, the Communist Party, now under the leadership of Mao Zedong (Mao Tse-tung), expanded rapidly during the war years because of its ability to win the support of people living in wretched conditions. Mao appealed mostly to the Chinese peasants with the simple slogan, "Land to the tiller." After the Japanese surrender, the Communists challenged the Nationalist government and gained control of mainland China in 1949. The defeated Nationalists were forced to flee to the island of Taiwan, which had only recently recovered from 50 years of Japanese colonial rule.

The new Communist regime called itself the People's Republic of China (PRC) and made Beijing (Peking) its capital. As chairman of the party, Mao Zedong, became the absolute ruler of China for nearly 30 years. During the first eight years, 1949–1957, the PRC made an alliance with the Soviet Union, intervened in the Korean War, and consolidated its control over the entire Chinese mainland, including the frontier regions of Inner Mongolia, Xinjiang, and Tibet. It restored peace in the country, pursued policies of economic rehabilitation and development, and carried out the first stage of its "socialist revolution" by eliminating rural landlords and urban capitalists.

Relative success in the PRC's domestic and foreign policies led Mao and his supporters in the party to pursue a more radical policy over the next 20 years to revolutionize China's

Mao Zedong, who became the leader of Communist China in 1949, at first appealed to the people by offering them land for farming. Over the course of his rule, however, he brutally tried to force the Chinese people to live according to a strict Communist philosophy. In the process, he caused terrible famine and almost destroyed the Chinese economy.

economic and social structure. Campaigns to launch the so-called Great Leap Forward in economic production and to set up People's Communes to incorporate all the rural peasant population, resulted in great disasters for the population. Human policy errors, combined with unfavorable natural conditions in

1959 and 1960, caused widespread famine in the country. An estimated 20 to 30 million people died during this time. The government parted ways from its Communist ally, the Soviet Union, and began to challenge both the United States and the Soviet Union for leadership in the world, particularly among underdeveloped countries. Chinese-Soviet relations reached a breaking point in 1969 when the two Communist giants fought a series of border clashes.

In 1966, Mao and his followers launched the Cultural Revolution with the avowed goal of establishing ideological purity for the Chinese people. It was an intense power struggle for Mao to purge his political rivals, real or imagined. Millions of people came under political persecution and both industrial and agricultural production suffered decline. The "ten disastrous years" of the Cultural Revolution ended only with Mao's death and the purge of the radicals in late 1976. Even before the Cultural Revolution came to an end, however, China made a major shift in foreign policy. It allowed U.S. President Richard Nixon to visit China in February 1972, and paved the way for the two countries to have normal diplomatic relations. China also began diplomatic relations with other Western powers and gained admission to the UN General Assembly, replacing Taiwan as a permanent member of the Security Council in 1971.

Post-Mao Period

A new leadership emerged in China after 1976 with the rise of the pragmatic Deng Xiaoping. While maintaining the monopoly of political power for the Communist Party, Deng's economic policy marked a sharp departure from Mao. He promoted reform in various sectors of the economy, promoted growth in foreign trade, and opened China for foreign investment. The success of the reform led to rapid economic growth in the 1980s and 1990s. China began to emerge as an economic powerhouse.

Such drastic change in the economy inevitably affected China's social and political structure. Rising discontent with official corruption and inflation, coupled with the growing demand for democracy, brought about massive student demonstrations at Tiananmen Square in Beijing during the spring and summer of 1989. The government treated this as a form of social protest and called out the military to suppress the demonstrators by force. The clash resulted in the death of hundreds of student demonstrators. Such disregard for human rights was widely condemned, and China was ostracized by the international community. However, China's leaders made no change to the policy of economic reform and opening to the outside world. Gradually, the nation's economic strength won back the favor of foreign countries.

Through the end of the twentieth century and the start of the twenty-first, China has pursued a good neighbor foreign policy. The fall of the Soviet Union in 1991 made it possible for China to restore cordial relations with Russia and other new states in central Asia. China has also won back its former territories through negotiations with Great Britain and Portugal; Hong Kong and Macau have returned to Chinese sovereignty. An unresolved problem for China is the continued existence of independent Taiwan, which is still outside of China's jurisdiction. This is in spite of growing ties in trade, investment, and cultural exchange between the two sides.

In recognition of the nation's new status in the world community, China's capital, Beijing, won the right to host the 2008 Summer Olympics, and the World Trade Organization finally extended China full membership. China is working to create a capitalist, market-oriented economy under a Communist framework of government. The imperial dynasties, which lasted over thousands of years, have quickly disappeared to give China a new orientation for the twenty-first century. Such a change would hardly have been contemplated just 50 years ago.

The Chinese have long enjoyed a colorful culture. These traditional drum dancers performed at a celebration of the Spring Festival, China's most important holiday, in 1997.

4

People and Culture

China contains one-fifth of the world's population. There are five times as many people in the country as in the United States, and the number of people is expected to climb.

Population

Three significant trends are likely to occur. The first trend is the aging of the population. By 2025, there will be 300 million people over age 60. It is projected that this age group will comprise 25 percent of the population. The second trend involves the urban population, which is expected to account for 55 percent of the total population. Thus, urban areas will continue to grow and become increasingly overcrowded. The central and eastern provinces of Henan, Shandong, and Sichuan are the most populated ones in the country. Each has approximately 90 million people. Chongqing, one of the four directly administered municipalities of China, located east of Sichuan province, is administratively the largest city in the world, with more than 30 million people. The immediate urban area has a population of approximately 4 million. China has 20 cities with populations of 5 million or more. By 2025, the urban population could reach 1 billion. A third expected trend is the movement of large numbers of rural people to the cities. It is these people who will form the bulk of the labor force needed for urban construction and expansion. China plans to expand some 600 towns into cities by the year 2011.

The population density of the country is 332 people per square mile (128 per square kilometer), but this figure needs regional interpretation. Each of 11 coastal provinces, from Guangxi in the south to Liaoning in the north, has a population density of 992 people per square mile (383 per square kilometer). For nine centrally located provinces, from Hunan in the south to Heilongjiang, including Inner Mongolia, in the north, the figure for each is 381 people per square mile (147 per square kilometer). Each of the remaining nine western region provinces has only 132 people per square mile (51 per square kilometer). The United States has approximately 75 per square mile (29 per square kilometer). Four municipalities—Shenyang, Tianjin, Chengdu, and Shanghai—have 90,650 people per square mile (35,000 per square kilometer). Chinese cities average 25,900 people per square mile (10,000 per square kilometer).

The spatial distribution of China's population is highly uneven and generally reflects the country's climatic patterns. Ninety-five percent of China's population resides in China Proper, the humid eastern region. This part of the country has 43 percent of the land area of China and most of the prime agricultural land. A closer look reveals that 80 percent of China's population lives in the four major river basins: the Northeast Plain, or the Liao-Songhua River Basin; the lower Huang He (Yellow River) Basin, or North China Plain; the upper and lower basins of the Chang Jiang (Yangtze River); and the Xi (West) and Pearl River basin of south China. Essentially, population densities are high along the coast, remain high along major river valleys westward, and gradually decrease farther inland. One exception is the inland Sichuan Basin of Chang Jiang. Estimated population density for the basin is 1,295 people per square mile (500 per square kilometer). In the far west, such as the Tibetan Plateau, there are fewer than 26 people per square mile (10 per square kilometer).

Though the Chinese population has grown steadily over the centuries, it increased dramatically during the first two decades of Communist rule. During the years of Mao's rule, population increased because birth control was thought to be anti-Communist, and Marxist philosophy encouraged more workers for the production of more economic goods. After Mao's death in 1976, family planning programs became strict, with the government promoting a rigid, one-child-family rule. Because of the one-child policy, China's projected growth rate for 2001–2050 is only 8 percent, compared with the expected 58 percent in India, which has no family planning policy. The Chinese government introduced a complex system of economic rewards to help achieve this one-child family. These included better living conditions, extra grain rations, and improved education and employment opportunities for the single child. Parents with two or more children faced fines and abortion threats, and would often abandon a girl child, who was less desirable in Chinese society. Traditions remain strong in some regions, though, and some families still want more than one child. As a general rule, most minority regions were excluded from the one-child policy.

Traditionally, Asian societies, including the Chinese, favor male children. This is more often the case in rural areas, where males are seen as more productive in agricultural work and for helping aging parents. The gender ratio varies, depending on region, from 105 to more than 125 boys for every 100 girls born. The world average is 105 boys to 100 girls. Obviously, a one-child society and a growing elderly population will mean a greater responsibility for both the government and the single child in the coming years. In China, 25 percent of the population is younger than 15, compared with the world figure of 30 percent.

Language

The language of China is the most widely spoken native tongue on earth. Although spoken dialects are different, the language shares the same writing system, with the exception of the scripts used by notable ethnic minorities, such as the Uighurs, Kazaks, Tibetans, and Mongols. Chinese written characters, or *hanzi*, allow two people who speak different dialects to be able to read each other's writing anyway. The spoken language is tonal. To pronounce a word properly, it is necessary to know not only the sounds of consonants and vowels, but also the four different tones of each. In the late 1950s, the government tried to simplify the language. It adopted the so-called *pinyin* system of standard Chinese. It is based on the pronunciation of the Chinese characters in northern Mandarin, and the Chinese spoken in the Beijing region.

Mandarin is the language of more than 800 million people around the world. It was only in 1956 that the government officially declared Mandarin the national, or common, language. It is the language used in government, schools, radio, television, and movies. It has become the language most frequently spoken in large urban centers.

Other major dialects are Cantonese (or Yue), spoken in southern China and Hong Kong; Wu, spoken in Shanghai and the nearby provinces of eastern China; and Min, spoken in southeast China. The Cantonese dialect is quite widespread. It is spoken by most Hong Kong residents, and is generally favored in Chinese communities overseas, including parts of Southeast Asia. Not all languages spoken in the country are dialects of Chinese. In the far west, the Uighurs and Tibetans speak languages that have little relationship to Chinese.

The great majority of Chinese have traveled little in the past, even within their own provinces. This lack of contact between the various regions has allowed diversity to develop. Thus, a wide variation of dialects exists even within provinces. For example, the Fujian dialect spoken in the coastal southeast area shows differences even between neighboring valleys. For the Mandarin-speaking people who live over a large area, it is not unusual to note dialect differences within distances of only 20 miles.

Ethnic Picture

China views itself as a thoroughly unified country. In large part, this is because the Han people, who account for 94 percent of the population, consider themselves the original Chinese people. They trace their lineage as far back as the Han Dynasty of 206 B.C. The Han live mainly in central and eastern Monsoon China. They are also scattered widely in north-central China and across the Tibetan Plateau. The Han account for more than 90 percent of the population in 19 of the 22 provinces.

Six percent of China's population, a total of more than 70 million people, belongs to 55 other ethnic groups. Fifteen of these minorities comprise 90 percent of the ethnic minority population. Some ethnic minorities have been given special treatment by the government through special administrative status. These minorities are officially recognized, and their language and customs are somewhat encouraged. At the highest political level of recognition, the government has established Autonomous Regions, or ARs. These ARs occupy the border areas of the country and are strategically located. There are currently five ARs, and each has a dominant minority.

In the far west, the Tibetans in Xizang and the Uighurs (the country's third largest minority) in Xinjiang, comprise two of the ARs. In the north, the Mongols in Inner Mongolia and the Hui (the country's second largest minority) in Ningxia comprise two other ARs. China's largest ethnic minority, the Zhuang in Guangxi, in the southwest, occupies the fifth AR.

It should be noted that three of the five ARs are located in large sparsely populated areas. These are China's frontier border regions and act as buffer zones for the country. Tibet is a buffer with Burma, India, Bhutan, and Nepal. Xinjiang is a buffer with Pakistan, Afghanistan, Tajikistan, Kyrgyzstan, Kazakhstan, and for 25 miles (40 kilometers) with Russia in the extreme northwest. Inner Mongolia is a buffer with Mongolia and Russia's far east.

The ARs guarantee political equality for minority groups. Yet in some cases, Chinese rule has been fiercely contested and conflicts have occurred with the central government. Two noted examples of such conflicts involve followers of the Dalai Lama in Tibet, and the discontentment of the millions of Muslim in Xinjiang Autonomous Region. At times, Uighur nationalists have taken very extreme protest measures by detonating bombs in Beijing and in the provincial capital, Urumqi.

The Chinese government has pressed the minority peoples to learn Mandarin Chinese if they want advancement and employment outside their immediate farming villages. Still, many minority groups prefer to maintain their distinct cultures, languages, and religions. This continually poses problems for the central government. Minority groups have created some complex situations, such as the 5 million Mongolians who are Lamaist Buddhists, and the 1.5 million Kazaks and the 9 million Uighurs in Xinjiang Autonomous Region who are Muslims. There are also 5 million Tibetan Buddhists, who belong to the Tibetan-Burmese language group. Many are loyal to the exiled Dalai Lama, who now lives in India.

Religion

Religious beliefs have profoundly influenced Chinese society from the earliest times. Confucianism, Taoism, and the assimilation of Buddhism from India helped to create the rich and unique system of thought that structured Chinese society for more than 2,000 years. The western religions of Islam and Christianity have made inroads over the last 1,300 years.

Chinese religion tends to be centered on the family and the local community. Taoism, founded in the sixth century B.C., promotes oneness and harmony with nature as a way of life. It can involve meditation, philosophical debate, and magic. Taoists deeply influenced Chinese arts, particularly painting and poetry. Related to Taoist ideas was the concept of *yin* and *yang*, the two opposing forces believed to be present in all nature. This was the idea of dark and light, or winter and summer.

Confucianism, more a philosophy than a religion, was also founded in the sixth century B.C. It is more involved with government workings and interpersonal relationships. One of its aims was to promote an ethical theory, placing emphasis on the dignity of the human being.

Buddhism, a foreign religion from India, stressed the idea of breaking the cycle of reincarnation and becoming an enlightened being, or *Bodhisattva*. Buddhist ideas have had a strong influence on Chinese culture, including art, architecture, poetry, and fiction. Followers of Buddhism have faced hardships. During the 1950s and the Cultural Revolution period of 1966–1976, thousands of Buddhist monasteries were shut. In Tibet alone, 6,000 monasteries were closed or destroyed, and only 25 were left.

The early Communist period was a time of anti-religious policies, and damage was inflicted on temples, monasteries, churches, mosques, and other religious buildings. In effect, religion was forced underground. The Communist leader Mao Zedong was elevated to a godlike status. His painted portrait, still prominently displayed at the entrance to the Forbidden City in Beijing, overlooks Tiananmen Square. His face is also pictured on the current 100 Yuan bill. Mao's influence was especially promoted during the Cultural Revolution. During this time, all books were devoted to Mao's thoughts and sayings. Among them was the Little Red Book, "Quotations of Chairman Mao." His portrait and statues are still found in public and private places. In early 1982, the government permitted greater religious freedom, but with strong controls. Prohibitions still persisted, especially in Tibet, where religious beliefs seemed to override central government policy. Recently, religious beliefs have begun to be practiced more openly. Yet the government still monitors activities closely.

Art and Popular Culture

Chinese culture has a rich artistic and intellectual heritage. The Chinese writing system dates back to the sixteenth century B.C. The Mao period dealt a vicious crackdown on all aspects of traditional Chinese culture, but conditions improved after Mao's death. A great array of historians, writers, poets, artists, and musicians made great contributions to the culture over the centuries. Such developments as calligraphy (literally meaning "beautiful writing"), painting, the care of gardens with miniature plants (bonsai), and the keeping of ornamental fish and birds, add to the richness of the culture. Elderly men gather in local parks to enjoy their caged birds and socialize.

Chinese culture is perhaps best shown by the tremendous variety of ideas and inventions that first appeared in China, which include the seismograph, the magnetic compass, gunpowder, paper, fireworks, porcelain, the suspension bridge, horse collar, and crank handle. Also from China came important medical discoveries, such as blood circulation, the thyroid hormone, smallpox immunology, acupuncture treatment, and deficiency diseases. The Chinese have always excelled at mathematics and many of their discoveries, such as the famous Chinese Remainder Theorem, are well known in the Western world. Generally, these Chinese ideas took five, ten, or even fifteen centuries to reach the Western world.

As Chinese society becomes more urban, traditional culture gives way to big-city life. Exposure to foreign influence, such as the fast food outlets of McDonald's and Kentucky Fried Chicken, drinks such as Pepsi and Coca Cola, and Western

One of China's most popular hobbies, especially among the elderly, is the keeping of decorative fish and birds. This Chinese man is shopping for a cage for his collection of ornamental birds.

music, fashions, and electronic gadgets, make for a changing culture. China has some 300 million cellphones and 90 million Internet users. Not all people favor these trends.

Family is still of major importance in Chinese culture. Each child lives in a very close, loving family structure. Generally,

Asian societies treat a young person up to about age eight as a Golden Child, who can do almost no wrong. Chinese society is also concerned about preserving hierarchy. Status in the family hierarchy is always foremost. The Chinese extended family encourages elderly parents to assist and influence family life. The effects of any disgrace touch even the extended family. The concept of honor, or "saving face," is important. As a result, disagreements are best settled in private.

Literature

Some Western writers suggest that there are certain difficulties in the enjoyment of Chinese literature and poetry. The Chinese language itself and translation into other languages are obvious barriers. One needs a vocabulary of more than 5,000 Chinese characters and an intimate understanding of Chinese history and literature dating back 2,500 years in order to appreciate the subtle nuances of what is written, particularly in Chinese classical poetry.

Even calligraphy, the art of freehand writing, requires extensive study and training. It takes a great deal of practice to write properly with a brush. This art is declining in China because specialized education is not widely available. Mass education and the use of modern pens also contribute to the decline.

Traditional Chinese literature is closely linked with Confucius. The recorded words of Confucius and his disciples, known as Classics, were considered a beneficial influence on society and the state. The Classics became the basis of education in traditional China.

A second category of traditional literature includes historical writings. Few other nations are as aware of their history as the Chinese are. The first general comprehensive history was compiled around 100 B.C. It contained political, social, economic, cultural, and geographic records from antiquity to the Han Dynasty. Later, historians focused on a particular dynastic period rather than general history. In total, 24 "official histories" were compiled by the end of the eighteenth century, unquestionably the most voluminous historical records of a nation.

The third category was made up of philosophical works. These were written by philosophers and thinkers of various schools of thought from ancient times to the early modern period.

The final category included literary works. Among these was a variety of collected works of poems, essays, literary treatises, and personal letters. Poetry, because of its creativity in expressing deep feelings, has always been the most appreciated and honored form of literature in China.

In traditional China, success depended on an educated person's mastery of these literary works. The Chinese civil service was based on a series of examinations that tested a candidate's knowledge of literature and ability in literary composition.

Later literary developments included drama, opera, and fiction. At first, these works were not considered respectable because the narrative and dialogue used everyday vocabulary, not the literary style needed to achieve social status. Eventually, such works became an accepted form of literature.

Society and Health

Health and food are important cultural concerns for the Chinese. Cities have Western medical facilities as well as traditional Chinese medical clinics. The traditional approach is a holistic one, which takes into account the entire body rather than a single complaint. There are many herbal remedies for a wide variety of ailments such as headaches, backaches, or colds. One widely practiced traditional technique is acupuncture. Needles are inserted at specific points on the body, each of which is thought to be linked through circulation to a particular organ.

Many Chinese maintain their health by the practice of *tai ji*. These exercises are practiced throughout China in public places in the early morning hours. They are generally followed by the elderly, who hope to maintain flexibility through slow graceful movements in a ballet-like dance.

One health issue facing China and other Asian societies is tobacco smoking. Since the 1980s, the amount of farmland in China used to grow tobacco has tripled. The country imports and exports tobacco on a large scale. Estimates are that one in every three cigarettes smoked in the world is smoked in China. By 2050, smoking will kill more than 8,000 people each day in China, most of them male. It has been said that the biggest cigarette company in the world is the Chinese government. Tobacco contributes up to 10 percent of total government revenues. The Chinese government is beginning to pay attention to the harmful effects of tobacco, and may take action to improve the situation.

The Chinese love of food is thought to be tied to health. China boasts one of the world's great cuisines. There is great variety and many regional cooking styles. The four major regional styles of Chinese cooking are northern (Beijing), southern (Cantonese), coastal (Shanghai), and inland (Sichuan). Vegetarian food is popular, with the ever-present rice dish. Rice and wheat flour are served in various forms—grain, noodles, or dumpling wrappers. Meats include pork, beef, fish, and fowl. Other foods on public display at restaurants include live snakes and seafood. Custom says that it is best to eat meat immediately after it is killed and cooked. In the past, lack of refrigeration led to this practice. Chinese youth today are opting more often for a Western-style diet, and over the years, have dramatically increased their consumption of meat.

Research suggests tea originated in China, or was introduced into China from Southeast Asia, some 2,000 years ago. Over the centuries, it became an integral part of the culture, and tea houses grew in large numbers. Tea brewing and drinking has become a ceremony in itself. The teacup in China is seen everywhere, like the coffee mug in North America. Many tea varieties reflect regional variations and preferences. Tea plantations are evident on the hillsides of southern China.

The Chinese are famous for their cuisine, which boasts a wide variety of dishes that are cooked with everything from pork to snakes. This man is cooking a cabbagelike vegtable, bok choy, to sell at an outdoor market.

Beer is a popular drink in the country, and rivals tea as a preferred beverage. The first brewery was established in coastal Qingdao, in Shandong province, by the Germans in the nineteenth century. The Tsingtao (Qingdao) label is the most popular. However, every province in China produces its own brand of beer. In some areas, beer is much cheaper than bottled water.

Chinese culture has endured for many centuries. It is rich, complex, unique, innovative, and enduring. Many Western observers have only a superficial knowledge and understanding of this culture, which will likely continue to flourish and deepen its historical roots, even in light of the strong influences of the Western world.

Although the Chinese government is still under the control of the nation's Communist Party, over the last few decades, political leaders, including these members of the National People's Congress, have begun to implement changes designed to help improve the Chinese economy.

5

Government

China has been recognized as a cultural unit for several thousand years. The exact limits of the Chinese Empire, however, have varied over the centuries. Control of the territory has sometimes been strong and widespread, but at other times, it has been weak and fragmented.

Traditional Government

The political organization of broad areas of China into a unified state dates from at least 1500 B.C. By then, Chinese society had developed a unique writing system, whose characteristics are still evident today. This early historical state was centered east of present-day Xian, and focused on the fertile soils of the middle Huang He Valley.

Throughout much of its long history, China has remained, in theory or in practice, a unified state. When the central government faced problems and disruptions, portions at the edges of the Chinese Empire often detached themselves and functioned as autonomous states, but the separation rarely lasted long. Even the emergence of local warlords who controlled large regions, as in the waning days of the Manchu Dynasty and the early years of the republic, could not prevent a return to the authority of a central government in China.

The Chinese generally divide their history based on the reigns of the various emperors of the ruling dynasties. If rulers were just, fair, and effective, they received widespread support and the authority to rule. Improper or ineffective rule or corruption, as shown by poor harvests or losses in battles, led them to lose this support. A new dynasty would then emerge. This principle first appeared during the Zhou Dynasty of 1050–256 B.C. The Zhou probably founded the mandate concept as a convenient way to justify their overthrow of the unpopular last ruler of Shang Dynasty.

The persistence of the Chinese Empire, however chaotic and fractured at times, was the most impressive feat of the Chinese political tradition. This empire was the equal of the Roman Empire, but was confined to Asia. Outside communications were less extended and less important, compared with the Roman Empire or the more modern Spanish Empire.

China has always been a strong, unified state headed by a single, powerful ruler, the emperor. Each dynasty functioned as a state, and could extract substantial revenues and labor from large populations. The bureaucratic system worked remarkably well and functioned without much major change for 2,000 years. From its founding in the second century B.C., to its fall in 1911, the system evolved in cycles, with deteriorations and restorations. New dynasties always looked to restore the system as it existed, with modifications, not to build a new order. The key thread for the Chinese state was the idea of a unified state with a strong central government led by a single ruler exercising direct control over local governments and neighboring lesser states. Deviations from the norm of a unified central imperial form of government were only evident in some five of the 20 centuries of China's political history.

The reach of the state was at its greatest in the eighteenth century A.D., during the middle years of the last dynasty. However, it was first unified as a nation-state under the rule of the first Qin emperor as far back as 221 B.C. It occupied about one-half of China today, north to the Great Wall, and westward from the capital Xian, to the border of Gansu Province. It also aspired to occupy areas as far south as Vietnam and as far east as Korea. The consolidation of the unified empire was made possible by the building of roads and canals, allowing expansion west and southwest.

Social order was maintained from the imperial throne down through bureaucratic officials, the scholar-gentry, to the commoners who included peasant-farmers, artisan-craftsmen, and merchants. There was a careful division of labor and a balance of power and authority for all sectors. All social classes were tied together by the ideas of Confucianism that stressed harmony in human relations for the community, resulting in social cohesion and order. From the Han Dynasty (206 B.C.–A.D. 220) onward, Confucianism was institutionalized as an ideological foundation of government, and prevailed in principle for 2,000 years. This was the beginning of the Chinese bureaucratic empire. Educated people studied for imperial examinations, and if they passed, they were granted government posts. Examinations stressed knowledge of Confucian classics, so the bureaucracy was governed by certain rites and rituals throughout the centuries. It was the best way to move up in society. It was based on knowledge and ability, and anyone could take the examinations.

The Chinese form of government over the centuries has functioned essentially as a massive bureaucracy. It operated as a political, administrative, and military machine that oversaw a controlled territory. At the top was the emperor. The emperor met with officials and advisors to make domestic and foreign policies, but was normally surrounded by only a small group of trusted advisors. A chancellor, two grand secretaries, and six ministers, including ministers of civil affairs, military affairs, and justice, were directly below the emperor. Local government included a series of administrators at various provincial and district levels.

Great attention was given to ritual and ceremony in conducting state business. These imperial institutions were designed to reflect a sense of awesome power and unapproachable remoteness from the people. The heavily guarded Forbidden City, or Imperial City, allowed the emperor to conduct daily affairs away from the prying eyes of common people. The Forbidden City itself symbolized the emperor's unique position at the center of the universe. This palatial complex is considered the incarnation of the celestial palace on earth.

The emperor always had special distinctions. No duplication of the emperor's dress was allowed. The emperor wrote in red, while his officials wrote in black. He alone faced south; his audience faced north. Only he used the special term, *Zhen*, meaning "I, the imperial majesty." Few people saw the emperor face-to-face. All imperial processions were conducted with an air of sacred splendor and dignified isolation. Imperial power was absolute, as long as the emperor seemed to govern with virtue. The bureaucracy was an important part of an emperor's rule because it implemented edicts throughout the territory. Traditional values and ritual symbolism followed from dynasty to dynasty. This is why stability re-emerged even after prolonged periods of chaos, conflict, and division.

The Ming emperors, 1368–1644, employed over 70,000 eunuchs in their bureaucracy. Eunuchs were castrated male attendants whose official job was to supervise daily business in the palace. They were used in Chinese courts for more than 2,000 years, but they reached the height of their power during the Ming Dynasty. The eunuchs were important links between the inner imperial world and the outer bureaucratic world. By the 1590s, they played a central role in the political life of China, collecting revenues in the provinces, using military guards to terrorize wealthy families, controlling food supplies, and even writing historical works. They were a distinctive part of the Chinese bureaucracy and had a great deal of political influence. After this period, they never again reached such numbers and their influence declined.

Money was an important part of dynastic rule. It was necessary in order to maintain an extensive bureaucracy and to keep a well-trained army that would defend and oversee a large empire. Taxation of the people was necessary. To lessen tax demands on people, it was important to produce an agricultural surplus, above the subsistence level. Money was to be used effectively. Dynasties often failed because of waste, extravagance, and heavy taxes. The power of an emperor could decline if local provincial bureaucrats gained local support and functioned independently. Any economic improvement, such as the Grand Canal of the seventh century or an upswing in rice production, was important in raising money to pay and feed the armies.

All dynasties had to maintain effective communication and transportation between the central government and the frontiers. Foot runners, horses, and boats were all part of a government courier system. Effective government rule required an efficient transportation system, including canals, bridges, and roads.

Maximum territorial control was first established under the Tang Dynasty, A.D. 618–907. It was impossible to control such a vast territory for long, however. Extension of territory always meant that loyal government officials and soldiers must be present in the outer territories to oversee government business. In times of peace, soldiers doubled as farmers. The soldier-farmer was key to keeping territory under control and united. The idea was to have self-sufficient garrisons, even in semiarid regions. Military settlements were walled for protection.

The physical character of China's boundaries allowed any given dynasty to maintain strong rule, assuming there was domestic tranquillity. The eastern seaboard, the rain forests of the southeast, and the western and southern mountain ranges were formidable barriers for any outside penetration of the territory. The only foreign intrusions came from the north, where open country allowed troops to move more easily. Historically, nomadic tribes, such as the Huns, Turks, Mongols, and Manchus, all invaded China from the north.

The rise of regional independent military power was always a concern for a dynasty. The central government always stressed the civilian side of imperial rule to keep the territory in check. Army personnel would be moved from one location to another. The idea was to prevent army generals from building local power and support, which might allow political fragmentations. At times, this constant change made the troops less effective and the dynasty more vulnerable to a new political regime or outside influences.

The Ming (1368–1644) and Qing (1644–1911) Dynasties showed great political skill in maintaining imperial unity for more than two centuries. Part of this success was achieved because of firearms. By rebuilding the Great Wall with fixed cannons, the Ming emperors also resisted invasions from the north for over a century and a half. The Grand Canal was used effectively to move military supplies north and south. Thus, the dynasty had a proven and effectively deployed army. The succeeding Qing Dynasty simply allowed the army to continue as before. This helped in the annexation of parts of Outer China and vast areas in the northwest.

It is important to note that dynastic rule for China had a strong foundation. The early Shang Dynasty, 1600–1050 B.C., and the Zhou Dynasty, 1050–256 B.C., lasted 1,400 years. After a short interlude of 50 years, the next dynasty, the Han, lasted another 400 years, from 206 B.C. to A.D. 220. Two thousand years of dynastic rule has strongly influenced more recent governments in

For thousands of years, China was run by a series of emperors who inherited their title from their fathers before them. The imperial system was finally ended in 1912, after Nationalists led by Dr. Sun Yat-Sen established a republic in China.

China. The present Communist form of government is in many ways similar to dynastic rule: There is one strong leader, a bureaucracy, and a large, well-trained army. Communism has been in place for only about 50 years, and it is already experiencing modifications and transitions. It is too early to tell what type of government structure will eventually emerge.

Contemporary Government

Today, the government of China is under the control of a single party, the Chinese Communist Party (CCP). A few other so-called democratic parties are allowed to exist, but are largely deprived of political power. The Communist Party, with a huge membership of over 68 million. It is the world's largest political organization and dominates the political, economic, and social life of the population. Those who lead the party formulate policies for the entire country. The party's role is not confined only to politics and ideology, but extends to all aspects of life, including industry, commerce, agriculture, education, social organization, and the military. The party controls all major appointments, and most leading officials are party members. Every major branch of the Chinese government is overseen by a unit of the party.

China's Communist Party is highly centralized. In theory, local party members elect representatives to form county party committees. Members of the county committees elect representatives to the provincial party committees. In turn, members of the provincial committees elect representatives to attend the National Party Congress, which meets once every five years. A Central Committee consists of up to 200 members and is elected by the Congress. It is responsible for policymaking and the approval of major personnel appointments. Because the large Central Committee meets only a few times a year, it elects a Political Bureau (Politburo) of two dozen members to take charge of important party affairs. The Politburo, and particularly its seven-member standing committee, contains the most powerful leaders of China. All members of the Politburo and most in the Central Committee serve concurrently in the top positions of the government. For instance, China's former top leader,

The former top leader Jiang Zemin was noted for his efforts to reform and improve the Chinese economy, in order to make China more competitive in world markets. In 2004, he was succeeded in all his political positions by Hu Jintao.

Jiang Zemin, 78, held all three top political positions. Recently, however, Hu Jintao, 62, an electrical engineer, was promoted to the same top three political positions: General Secretary, President of PRC, and Chair of the Central Military

Commission. The General Secretary of the party is also chair at meetings of the Politburo and the Central Committee.

The Chinese government is a pyramid bureaucratic organization parallel to the party structure. The central government, located at the capital of Beijing, is headed by a president. The executive branch of the government is the State Council, which is headed by a premier and several vice premiers. The State Council consists of more than 30 ministries and commissions. It is similar in size to the cabinet of a country like the United States, Russia, or India. A court system headed by the Supreme Court forms the judicial branch of the government. A National People's Congress with thousands of members chosen by electoral districts throughout the country, and nationwide professional associations, meets once a year at the capital. In theory, it has the power to approve major legislation and personnel appointments made by the State Council.

China is divided into 23 provinces, including Taiwan; five Autonomous Regions (ARs); and four directly administered municipalities of Beijing, Shanghai, Tianjin, and Chongqing. Taiwan is ruled by a separate Nationalist government and is outside of China's jurisdiction. These units form the first tier of China's regional government. Many provinces have existed in roughly their present shape for hundreds of years.

Each province or region is divided into counties, which number more than 2,000 in the entire nation. These form the second tier of regional government. On average, a county may have a population from half a million to more than one million.

The smallest local government unit in rural areas is the village, and in urban areas, the district. China is at the beginning of an effort to implement an electoral process for the selection of village and district officials.

Until recently, Hong Kong and Macau were colonies of Great Britain and Portugal. Hong Kong reverted to China's sovereignty on July 1, 1997, and Macau followed suit on

December 31, 1999. These two former Western colonies are accorded special status by the Chinese government because of their different historical background and socioeconomic conditions. Each was made a Special Administrative Region (SAR), and is run by a government elected by the region's population. China promised not to alter the status of these two districts for 50 years. The Chinese government offered these generous terms to try to woo people in Taiwan, in the hope of an eventual reunion of the island with mainland China.

Despite the rapid pace at which China has modernized its economy over the last few decades, in more remote areas of the enormous nation, the people are still tied to old-fashioned working methods. This man is using a traditional shoulder yoke to carry large loads of food.

Economy

China has a very rapidly growing economy, which has led to sharp increases in energy demand. The transformation of the economy, over the last quarter of a century, has been nothing short of a miracle. Communism, at times, has appeared to be secondary on the government's priority list.

This economic development has affected various parts of the country differently. Urban coastal cities and their surrounding areas, particularly in the south, have experienced more rapid economic growth than any other part of the country. In the 1990s, China achieved an exceptional average annual economic growth rate of more than 10 percent. In 2003, this figure remains at 9 to 9.5 percent. In the 1980s, the legacy of a state-run economy was reviewed by the central government and deficiencies were identified. The principle of state ownership of land was considered sacred in China.

In 1992, however, the government allowed limited property speculation in urban areas. This increased the price of land for real estate development. A mixed economy has since developed, with a combination of state-owned and private firms. The government has encouraged foreign investment, and since the 1980s, has allowed special economic zones to emerge. Now foreign investors receive preferential treatment for investing in these zones.

Daily life for people in Beijing or Shanghai is very different from the lives of the peasants in nearby rural Shandong province or people in the far western province of Xinjiang. Conditions for Chinese peasant farmers have not improved dramatically with the new economic direction. There is an increasing gulf between those who are very poor and the emerging rich of the population. Social unrest in the poorer western provinces, the break-up of financially troubled state-owned enterprises, and the movement of rural people to large cities have caused problems of unemployment and social unrest. World Bank figures suggest over 45 percent of the Chinese population live below the international poverty line of $2 US a day.

China is having a hard time accommodating Communist principles and capitalist business interests. Much of the Chinese economy is still controlled by large state-owned enterprises. Many are inefficient and unprofitable. An immediate priority for the government is to restructure some of these enterprises to private and more efficient operation.

Chairman Mao Period

During the 30 years after the founding of the Communist government in 1949, Chinese economic development was virtually halted by a series of political movements. The Great Leap Forward and the Cultural Revolution through the 1950s and 1960s severely damaged China's economic structure. The Great Leap Forward was Mao's idea to establish communes and to increase industrial production dramatically. Throughout the country, people were forced to make steel, even from household items, in simple backyard furnaces. The results were disastrous. The steel produced was poor quality and agricultural production fell severely when crops were neglected. Famine resulted.

The Cultural Revolution, 1966–1976, was Mao's scheme to punish his critics. Armies of young students organized into Red Guard units were encouraged to create a totally new China. Intellectuals, teachers, and parents were all targets of humiliation, and historic buildings and museums were destroyed. Students took over the universities and disrupted teaching programs, while teachers and intellectuals were sent to work on farms. It was a time of total disorder, confusion, and chaos for all Chinese.

The disruption in the country ceased only after Mao's death in 1976. The regime of Deng Xiaoping introduced reforms. Since 1982, recovery has been remarkable. The real economic growth rate exceeded the official target, which had been set at 8 percent.

Post-Mao Advances

A key part of this economic upswing began in 1978. This was a plan for renewal and development. Known as the Four Modernizations, the plan stressed new approaches for agriculture, industry, science and technology, and national defense. Foreign investment was encouraged. Thousands of students were allowed to study overseas. The new government promoted light industry, led by textile and food-processing enterprises. This was the beginning of the production of significant quantities of consumer goods in China, many of which were not made there before 1978.

The 1980s saw China make remarkable gains in industrial and agricultural production. Small-scale private enterprises were encouraged in both rural and urban areas.

China appeared to be on the threshold of new economic relations with other countries. In June 1989, however, the brutal suppression of student demonstrations in Tiananmen Square in Beijing caused the world to impose temporary economic sanctions on China. Tourism fell and the economy declined. Still, economic reform continued. One year later, the United States renewed China's most-favored-nation status and economic sanctions were soon lifted by other western governments.

Agriculture

When China introduced its economic reform policies in 1978, agriculture, as in the past, was a cornerstone. Farming in China started some 8,000 years ago. An old Chinese saying states, "Food is the Heaven of the People." Thus, agricultural production largely focused on food crops, rather than crops for industrial purposes or animal feed. Traditional agricultural regions include the river valleys in western China, where wheat, barley, and maize are grown; the far south and southeast, where rice paddies dominate; and the North China Plain, home of winter wheat. Important secondary crops include vegetables, melons, potatoes, sweet potatoes, peanuts, and soybeans. In selected regions, there are apple and pear orchards, and in the south, citrus fruits are grown. Fish ponds are also important, especially in the Chang Jiang and Pearl River lake district.

Sericulture

The oldest and most famous traditional handicraft industry in China is sericulture, the raising of silkworms for the production of raw silk. Even today, it is a large textile industry, second only to cotton. There is speculation that sericulture began as far back as 2600 B.C. Silk was the basis for China's earliest international trade. Mulberry plantations are found from Manchuria to the Hong Kong area. Mulberry leaves are the preferred food of the silkworm. The most famous plantations are in the Suzhou area, near Shanghai, and Zhejiang province. Other noteworthy areas are in Guangdong and Sichuan provinces. In the north, some silkworms feed on oak tree leaves, and their silk is a coarser type. There are many regional specialties of raw silk, depending on local climate and soil condition. Most of the 45,000 tons of silk produced each year is exported. China has captured 90 percent of the world market for raw silk, and 30 to 40 percent for finished silk.

Tea

The tea plant has long been of great agricultural importance for China. Its origins trace to the second century A.D. Grown in plantations in the uplands of central China and the ranges of the coastal provinces, tea is also important in the interior Sichuan province. Green tea accounts for 45 percent of tea production. Black and brick tea comprise another 45 percent, and wulung, chrysanthemum, and jasmine tea are other varieties.

Rice, Wheat, Millet, Cotton

In any economic picture of China, subsistence crops are far more important than commercial products. Some 55 to 60 percent of the cultivated farmland is occupied by the three great food crops—rice, wheat, and millet. Rice, accounting for up to 50 percent of food production, is grown in central and south China. Wheat and millet are grown in north China.

China is the world's top producer of rice, wheat, barley, millet, and sweet potatoes. It is the second largest producer of corn, sorghum, potatoes, and soybeans. The country accounts for 15 percent of global cotton production, the most in the world. Cotton is mainly produced along the middle and lower reaches of the Huang He and Chang Jiang valleys. The southern and southeast coasts are productive areas for sugar cane, the other major non-food crop.

These rice fields in south China are a common symbol of Chinese culture and economy, especially since rice is such an important component in so many Chinese dishes. China is, in fact, the world's largest producer of rice, but it is also a major producer of other goods, including wheat, millet, barley, and sweet potatoes.

Animal Husbandry

Animal husbandry, like farming, has a long history in China. Meat production of pork, beef, and mutton reaches

25 to 30 million tons a year and puts China first in the world. Yet this is still only 20 percent of China's total agricultural output.

China is the largest producer of swine in the world. One out of every two pigs is raised in China. Sichuan province has almost three times as many pigs as the American Midwest hog belt. The pig has long been part of the traditional Chinese diet and is valued not only for pork, but for its mythical existence. It is one of the 12 animals comprising the Chinese zodiac.

Fishing

Historical records show artificial fish ponds in China as early as 1142 B.C. Coastal fishing dates to the eighth century B.C. Since 1950, fish production has developed rapidly, but is a very weak part of China's agriculture, amounting to less than 2 percent of the value of total agricultural production. Though there are many fish farms, the sea is the main source of fish for China.

Future Challenges

China's agriculture faces serious problems. It functions at a low level of technology. There is still a traditional emphasis on farming, to the exclusion of fishing, forest enterprises, and animal husbandry. Unfavorable physical conditions, such as rainfall and temperature changes, play havoc with the stability of production. The demands of 1.3 billion people leave little room for error from year to year. Famines have occurred. The most pressing problem is the constant demand on the land caused by overuse and misuse, and urban growth. Land use planning with an emphasis on regional specialties and modern techniques, such as growing plants in water without soil (hydroponics), might help end these problems.

Another current problem in China is unemployment, as rural workers migrate to big cities. Unofficial estimates place the number of unemployed in urban areas at 20 million people. The employment picture changed dramatically from 1950 to 2000. In 1950, agriculture accounted for 84 percent of workers, service industries 9 percent, and manufacturing 7 percent. Recently, the figures were: agriculture and forestry 50 percent, service industries 28 percent, and manufacturing industries 22 percent. Out of China's total work force, 70 percent are rural workers. Most are employed in farming, forestry, fishing, and animal husbandry, or work on state-owned farms.

Agriculture has a declining share of the country's gross domestic product, accounting for some 18 percent of the GDP. This compares to almost 50 percent for industry. China's new leaders bring with them a new perception of China as a world economic giant with a vast potential market for foreigners. Yet agricultural production must be maintained at a meaningful level if China is to meet its future food needs, considering that the population is projected to be 1.8 billion by 2030. Irrigated land, now covering some 203,089 square miles (526,000 square kilometers), will also be an important part of any future agricultural picture. Nearly half of China's cultivated land is irrigated. There is fear of a water shortage, as demands for water become greater.

Industry

In 1949, the industrial sector in China employed only one percent of the work force. In the 1950s, as the Communist revolution swept through China, almost all industry became state-owned, and agriculture was organized into collectives. The economy was directed by the central government, following goals outlined in five-year plans. Industrial activity was concentrated in regions along the coast, with most industry found in former treaty ports, and major river valleys. Another region of expansion was the northeast (Manchuria), where Russian and Japanese colonial interests developed heavy industries based on local iron ore and coal resources. This early industrial development stressed heavy industry, such as iron and steel, chemicals, electric power, and textiles.

During the 1950s, much of China's industrial activity was related to building military items. Medium-sized facilities were based on regional self-sufficiency and the use of local resources. Such operations were located toward the interior, away from the more vulnerable coastline. Smaller industries were directed by the central government to produce agricultural equipment and building supplies. The idea was to promote greater peasant productivity and self-sufficiency for the communes.

Modern Development

China's industry has shown dramatic growth, especially after Deng Xiaoping took power in 1978. Despite the Cultural Revolution, policy struggles and reversals, civil conflicts, and dislocation of people, China managed to emerge on an even keel by the year 2000. It brought its population explosion under control with the one-child policy; it is improving city roads, bridges, and sewers; it has strengthened its military; and it has become a world player. Its trade, both regional and global, is growing.

Resources and Energy

The prosperous east coast provinces, with high-technology industries and foreign investment, are a major part of resource development. The central provinces have an emphasis on energy production, heavy industry, and agricultural processing. The far western provinces, because of their rich natural resources, concentrate on minerals, animal grazing, and agriculture where water is available. China's future construction projects involve developing a national transportation network to help long-distance economic exchange.

State-owned enterprises still control most commodities, including steel, coal, oil, and electric power. Prices are kept artificially low to maximize industrial profits. In the coming years, however, the central government is expected to raise prices to better reflect the world-market level.

China is rich in energy sources, such as coal, petroleum, natural gas, hydroelectric power, and nuclear energy. Coal produces 70 to 75 percent of the energy in China. It is the largest producer and consumer of coal in the world. Oil accounts for 20 percent, followed by hydropower at 7 percent and natural gas at 2 percent. Nuclear power is expected to generate 3 percent of electricity by 2010.

Coal

China produces 1.5 billion tons of coal yearly and ranks first in the world. It consumes about 25 percent of the world total and can still export, mainly to South Korea and Japan. Coal mines are found throughout China, but are concentrated in Shanxi province on the Loess Plateau. Another significant area is the Shanxi, Shaanxi, and Inner Mongolia borderland, which is referred to as China's "Black Triangle." Recently, large-scale open-pit mines have been opened. China is increasingly open to foreign investment in the coal industry, and has expressed interest in coal liquefaction technology and coal bed methane production.

A major environmental problem for China is the burning of coal. Heavy use of unwashed soft (bituminous) coal is the main source of air pollution. Sulphur dioxide and smoke dust are major pollutants. Power plants, factories, automobiles, homes, and trains also contribute. Acid rain affects up to 40 percent of the country. The World Bank recently reported that 13 of the world's 20 most air-polluted cities are in China. These are Tianjin, Chongqing, Shenyang, Zhengzhou, Jinan, Lanzhou, Beijing, Taiyan, Chengdu, Anshan, Nanchang, Wuhan, and Harbin. Further, the Chinese government has admitted that air is polluted in over 225 of its cities. It adds that water is also polluted in 7 major watersheds in the country.

China produces huge amounts of coal each year–more than any other nation–both for its own use and for export to other countries. Most coal mines, like the one in which this worker is employed, are still owned by the government, however China is becoming more willing to allow foreign investors to take an interest in mining. Although China makes money from the production of coal, pollution caused by the burning of coal as fuel has become a terrible problem in recent years.

Another concern is coal mining itself. It is exceptionally dangerous in China. Some reports indicate 14 coal miners die every day. Explosions, underground floods, negligence, poor ventilation, and lack of safety equipment are some of the causes. Conditions are not likely to improve as demand for power is fast increasing.

Oil

China has made great strides in oil production. It is the fifth largest producer in the world. Proven oil reserves for China are ranked ninth in the world, similar to the United States, with up to 24 billion barrels. The country, which is the third largest consumer in the world, became self-sufficient in 1965. This was, in part, because of a large oil field discovered in the northern Northeast China Plain. The famous Daqing field accounts for one-third of China's oil production. Other fields are in Xinjiang province, coastal Shandong and Liaoning provinces, and the coastal continental shelf of the Yellow, East and South China seas. China and five other South Asian countries are in dispute over the Spratly Islands in the South China Sea because of potential offshore oil and gas reserves. China and Japan also dispute claims over parts of the East China Sea for offshore resources.

Natural Gas

Historically, natural gas has not been a major fuel in China. Traditionally, natural gas was used as a feedstock for fertilizer plants. Little was used to generate electricity. The best reserves are in the Tarim Basin in Xinjiang province in western China. Domestic reserves look promising, but require a huge investment for a west to east pipeline to service east coast cities. For example, by 2005, a multibillion dollar gas pipeline, stretching 2,485 miles (4,000 kilometers) from Shanghai to Xinjiang and perhaps into Central Asia, will be a great advance as China moves to more reliance on gas.

Other domestic gas deposits in Shanxi province could more easily link to consumers in Beijing and northeast Hebei and Shandong provinces. Sichuan province in the southwest could supply consumers in Hubei and Hunan provinces in central China. Guangdong province uses imported liquified natural gas for power generation. Plans call for foreign investment to help build China's first liquified natural gas import terminal here.

Hydroelectric Energy

China, as a mountainous country, has hundreds of rivers. At the time of the Communist revolution in 1949, it had only 8 dams. Today, it is the world's leader, with more than 19,000 dams. However, only 20 of these have a capacity of 1,000 megawatts or more. Traditionally, dams were used for irrigation and flood control. Now the country has great potential for hydroelectric power. It ranks second in the world for water power potential and fourth in the world for hydroelectric power produced.

One spectacular dam that will play a prominent role in China's near future is the Three Gorges Dam. The 17-year project is set for completion in 2009. It is located on the middle Chang Jiang (Yangtze River) near Yichang city, Hubei province, and will be the largest dam in the world. It is expected to generate 84 billion kilowatt-hours of electricity a year. The reservoir is 375 miles (604 kilometers) long, creating a lake the size of Lake Superior, 32,000 square miles (83,000 square kilometers). It will be 575 feet (175 meters) deep, with an average width of 3,600 feet (1,100 meters). The wall holding the water back will be 600 feet (185 meters) high. The dam is expected to provide 10 percent of China's power needs, yet because of its immense size and the disruption of land and people, this dam is surrounded by controversy both within and outside of China.

The project will submerge 1,711 villages, 127 cities and towns, and 1,600 factories. More than 600,000 acres (242,810 hectares) of farmland will be lost. Up to 2 million people will be displaced. This will allow greater flood control for the river valley and give ocean-going cargo ships and passenger liners direct upstream access to Chongqing. This heavily populated city, situated 1,333 miles (2,145 kilometers) from Shanghai, will become the world's largest inland port.

Another project is scheduled for the upper portion of the Huang He, involving the western provinces of Shaanxi, Gansu, and Qinghai. Twenty-five generating stations are planned.

Modern industries in China are very dependent upon energy sources such as hydroelectricity, coal, petroleum, and natural gas. Important parts of the economy that depend on these include 14 major iron and steel plants in the central and eastern parts of the country including the largest at Anshan; chemical industries, including those that make chemical fertilizers and organic chemicals, clustered along the coastal areas; and construction material industries of all sizes, scattered throughout the country. Engineering industries are highly diverse. They produce tools, trains, trucks, bicycles, airplanes, and machinery. Centers include numerous east coast cities, and the inland cities of Wuhan, Xian, and Chongqing. Light manufacturing industries include paper and porcelain-making. A traditional cottage industry, still evident today, is gathering reeds along the Chang Jiang and its tributaries for the making of so-called rice paper. Rice paper is used for traditional painting and calligraphy.

Transportation

Road, rail, and air networks are important for the country. Railway lines total some 43,532 miles (70,058 kilometers), but only 11,600 miles (18,668 kilometers) are electrified. The U.S. rail network, by contrast, reaches over 141,961 miles (228,464 kilometers). China's international rail links reach Pyongyang, North Korea, Ulan Bator, Mongolia, and Moscow. Links to Xinjiang province in the far west, from the east coast, use the interior capital city Lanzhou, of Gansu Province, as an east-west hub. A $4 billion rail extension is planned for completion by 2006 to link Golmud city in Qinghai Province to Lhasa in

Tibet. Planners face the permafrost challenge and of reaching heights of 3.1 miles (5,000 meters). Thus, part of this line will become the world's highest railway.The north-south domestic links include Beijing to Guangzhou and Hong Kong in the south, and Beijing to Harbin, capital city of Heilongjiang province in Manchuria.

In 1945, China only had 50,000 miles (80,500 kilometers) of roads. Today, that figure is close to 1.2 million miles (1.9 million kilometers). Highway construction in China is proceeding very rapidly. At the moment, the Beijing-Shanghai expressway is the longest in China, stretching over 700 miles (1,130 kilometers). The country has about 12,427 miles (20,000 kilometers) of expressways. Many four-lane highways exist in the country, usually near large cities, such as Beijing, Shanghai, and Guangzhou. Freeways, elevated highways, and modern bridges were built since the early 1990s as cities expanded and modernized. Freight movement is increasingly important for China's roads. Even so, more freight is carried on inland and coastal waterways than by either rail or road.

Private car ownership is limited, and bicycles outnumber cars by at least 250 to one. Bicycles are a vital means of daily transportation everywhere. One in every three bicycles in the world is in China. There are 2 million privately owned cars in China and this figure is increasing daily. Western car manufacturers are targeting China. They see the potential for a vast car market, since only 0.1 percent of the Chinese now owns a car. Companies and taxis account for 70 percent of car purchases. But in 2004, China expects to produce 5 million cars, making the country third behind Japan and the United States. By 2010, China's car production forecast is for 10 million.

Taxis and public buses are abundant in all cities. Taxis have dramatically increased their ridership in recent years. Road transportation is increasing at the expense of water and rail transportation. The increase was over 90 percent in the ten-year period between 1990 and 2000.

Aircraft manufacture is becoming a noted industry in China. In Xian, the modern A320 plane is being built for domestic and international use. Airplanes fly more than 1000 domestic routes, 18 regional, and 85 international. The international routes reach 60 overseas cities in 40 countries. There is large air freight demand to and from North America, Europe, parts of Southeast Asia, and Japan. Air China flies overseas, and 15 airlines, including China East, China Southern, and Dragon Air, handle domestic flights.

In recent years, the government has realized the enormous economic advantages of promoting travel within China. The economic benefits are obvious when more than one billion people are on the move within the country. The improvement of domestic air travel, mainly using A320 planes, the promotion of efficient train operations, and even the extension of holiday time for workers are examples of government initiatives to encourage travel. By 2020, China will be the fourth-largest source of outbound tourism in the world, sending 100 million tourists abroad.

Special Economic Zones

The creation of special economic zones along the east coast allowed a significant break from a true Communist economy. As early as 1981, Shenzhen, adjacent to Hong Kong, was the first region to have greater economic freedom and favorable terms for foreign investment. Once a small sleepy peasant village with rice paddy cultivation, it was transformed into a landscape with skyscrapers, a new railway station, and airlinks to the rest of China and beyond. Shenzhen's urban population skyrocketed. Over a 20-year period, the population zoomed from 20,000 to 4 million. This is an example of the phenomenal growth rate for some of these special economic zones.

Rapid development swept the coast from Zhuhai, near

Macau, north to Shantou, Xiamen, and Shanghai's Pudong district. Later, part of Hainan Island off China's far southern coast, fronting the Gulf of Tonkin, was granted special economic status. It gave China closer access to all of Southeast Asia, when compared with economic rivals Japan and Taiwan.

China offers a huge domestic market, plentiful cheap labor, few restrictions on work conditions, and an authoritarian government that leans toward market reform. Deng Xiaoping's government reforms set the stage. However, the effect was confined to the coast so it would cause as little disruption as possible for interior China initially. Six eastern coastal provinces, particularly Guangdong, have greatly benefited, but at the expense of other parts of China. A quarter of all China's registered foreign enterprises are in Guangdong province. Manufacturing and real estate account for over 70 percent of all direct foreign investment. Hong Kong is by far the top investor in China. Japan, United States, and Taiwan follow.

In 1984, China started 14 east coast cities. These so-called open cities gave preferential treatment to foreign investors. Most of the cities had once been treaty ports, with a history of foreign trade. For example, the city of Qingdao has a central European atmosphere, with well-preserved old German architecture. The cities were selected based on size, links to overseas Chinese, established transport and industrial development, and the availability of local labor and talent. They stretch from Dalian city in Liaoning province in the north, to Beihai city in southern Guangxi province.

The east coast provinces, with their special economic centers, are most favored for the promotion of high technology and consumer goods manufacturing, service sector expansion, export-oriented production, and, of course, continued foreign investment.

Trade Patterns

China's more liberalized economic attitude has had an impact on the country's trade patterns. Ten years ago, China was fifteenth among the largest trading nations in the world. Now, including Hong Kong, it is ranked third. Trade is vital to China's future economic development, and it is the eastern coastal provinces that will gain even more advantages.

Japan is China's largest trading partner at $133.6 billion. Imports from Japan amount to $74.2 billion, exports are $59.4 billion. The United States is second at $126.5 billion with $33.9 billion in imports and $92.6 billion in exports. Third is Hong Kong at $87.4 billion. China treats Hong Kong as a separate trading partner. South Korea, Germany, and Taiwan are China's other major trading partners. International trade for China is rising from approximately $475 billion in 2000 to over $700 billion in 2004. The country is second to the United States as the world's leading recipient of direct foreign investment.

Manufactured goods account for 85 percent of China's trade. Exports include manufactured goods, clothing, footwear, electrical appliances, telecommunications equipment, transport equipment, and food and livestock products. China has become the chief supplier of low-cost consumer goods to the United States. Imports include machinery and transport equipment, basic goods, chemicals, and crude materials.

China is the fastest-growing export market for the United States. The trade balance with the United States was in China's favor in 1998 and growing. The Chinese government is continually trying to make concessions to prevent this imbalance from getting worse. Trade with the United States is never simple. It may be influenced by long-standing issues, such as human rights abuses, theft of nuclear information, military weapons and build-up, and the status of Taiwan. In 2000, China concluded a comprehensive trade agreement with the

United States, which paved the way for China's entry into the World Trade Organization in 2001. China agreed to open most sectors of its economy to increased foreign participation, allowing concessions in the energy sector, including petroleum.

China is fast becoming the hungry dragon. The country is demanding over 20 percent for each of these world commodities: crude steel, aluminum, copper, and zinc. It seeks to purchase 10 to 20 percent of each of the following world commodities: stainless steel, lead, and nickel. Further, in 2003, China took the following as a percentage of global totals: pork 51 percent, cement 40 percent, cigarettes 35 percent, steel 27 percent, televisions 23 percent, and cell phones (by number of users) 20 percent. Conversely, the country produces the following as a percentage of world totals: vegetables 76 percent, cashmere 70 percent , DVD players 60 percent, cameras 50 percent, winter clothes 50 percent, televisions 30 percent, personal computers 21 percent, and refrigerators 20 percent.

Monetary Policy

The exchange rate for China's official currency, the Yuan, was allowed to float on the international money market in 1992. At the time, one U.S. dollar was worth five Yuan. The exchange rate became one U.S. dollar to 8 Yuan in 1997, and the rate has remained stable since then.

Nowhere can the new direction of the Chinese economy be seen more dramatically than in Chengdu, capital city of inland Sichuan province. Once a home to poets and rulers alike, the city now boasts a stock market. Several thousand people gather there daily, after paying an entrance fee, to trade company shares. The central government does not recognize this activity and local authorities ignore it. Stock exchanges also exist in Hong Kong, Shanghai, and Shenzhen. People like to invest their money in stocks and they frequent the local stock trading offices. These are usually identified by the large number of bicycles parked outside, as evident in Hangzhou and other cities.

Current Challenges

Environmental problems continue to impact the country. Rapid industrialization and economic expansion, the unprecedented growth of Chinese cities, and a heavy reliance on the burning of coal contribute greatly to the severity of the problem. Air, water, and soil—the basic components of the environment—fare badly in China.

Rapid economic growth for China poses another serious problem. The demand for electricity is expected to double over the next 15 years. Now everyone wants a television and an air conditioner. As coal-fired plants emit pollution into the air, China's consumption of coal is projected to double by 2020. Skyrocketing demand for electricity could bring brownouts and blackouts. China's economic engine also could face serious interruptions.

One problem that accompanied China's new economic reforms in recent years was an increase in official corruption. A new get-rich-quick attitude developed at all levels, both in the government and the private sector.

Rapid economic development for China over the last several decades has created regional disparities. The latest UN China National Human Development Report of 2002 ranked all provinces and the autonomous regions and municipalities, based on an index that factored in lifespan, education, and income. Shanghai, Beijing and other coastal areas such as Tianjin, Guangdong, Liaoning, Zhejiang, Jiangsu, and Fujian ranked very high, numbers 1 through 8. Other coastal areas ranked moderately high, including Hainan Island. The lowest part of the scale, 25 to 31, were the island provinces of Sichuan, Ningxia, Yunnan, Gansu, Qinghai, Guizhou, and Tibet.

When China, in its entirety, is compared to the outside

world, using the same 2002 UN Human Development index, the country ranked 94 out of 177 countries. Norway was first, the United States was eighth, and Hong Kong was twenty-third.

If such trends continue, it is likely that regional differences in the quality of life inside China will only magnify in the coming years. Government mechanisms are necessary to redistribute the anticipated prosperity of the eastern coastal provinces, and to filter future development increasingly to the central and western provinces.

China is an interesting nation that combines the skyscrapers and advanced technology of today's modern economy with the traditional culture and farming practices of ancient times. China's changing economy in recent decades has helped cities grow into sophisticated centers of international commerce, such as Shanghai's business district, which is seen here.

7

Regional Contrasts

Geography has always been a subject of interest for Chinese scholars. A classic book written in the fifth century B.C., divided China into nine regions. For each region, an inventory detailed mountains, rivers, lakes, swamps, soils, and chief economic products. Another ancient work classified China's land into five major types. A later book, written in the third century B.C., divided land into first- and second-level types based on physical characteristics, such as hills, soils, and surface materials. Other geography studies were compiled and published over the centuries. Many of these were regional geography studies of the provinces and counties. More than 9,000 regional studies have been completed by Chinese scholars. Recent works studied regions and described resource and agricultural potential.

The Chinese Academy of Science has compiled separate 8- and 12-volume works that put China into regions, using physical characteristics and land types. One can divide China into six easily identifiable regions: Northeast; North China Plain; Chang Jiang Drainage Area; Subtropical South China; Inner Mongolia-Xinjian Steppelands; and the Tibetan Plateau.

Northeast

The Northeast, also known as Manchuria, encompasses the three provinces of Liaoning, Jilin, and Heilongjiang. It is considered a peripheral part of the country because of its very harsh winter climate, its isolation, and before 1900, its sparse settlement. Yet it has always been rich in natural resources, such as coal and iron ore, and it has available, though limited, agricultural land. It is strategically very important. It forms river boundaries with Siberia and North Korea. People from this region ruled China for the last 300 years of its imperial history. In the recent past, it has been coveted and occupied by Russia and Japan. With an area of 308,880 square miles (800,000 square kilometers), it has a population of 120 million. This region will continue to be important for China, given its nearness to significant Asian countries and the open Yellow Sea.

In the 1950s, the Communist government devoted considerable attention to this part of the country. The discovery of oil at Daqing focused even more attention here. Located 50 miles (80 kilometers) northwest of Harbin, capital city of Heilongjiang province, it is the largest oil field in China. The unique heavy industrial character of the area is evident in Anshan city, in Liaoning province, which produces 20 percent of China's steel. This makes the harbor at Dalian, at the southern tip of Liaodong Peninsula, the only important port for this northeast corner of China. The maritime climate makes this deep-water port ice-free. An extensive rail network helped in the development of this region as well. The Russians built part of the Manchurian railway through Harbin, linking Siberia to their far eastern naval port of Vladivostok.

Historically, the region was referred to as a great northern wilderness, with considerable mountain forest and open plains. Farming was introduced in the third century A.D. Over the centuries, many people have made the Northeast their home. The best known are the Manchus, the indigenous people of Manchuria. They developed an economy based on hunting and fishing. Gradually, the region became China's northern granary.

North China Plain

No region is more important to China than the Huang He (Yellow River) basin, which forms part of the strategic North China Plain. This is an area of over 386,100 square miles (one million square kilometers), with a population of 300 million. It includes Beijing and Tianjin on the northern margin, and the provinces of Hebei, Shangong, Shanxi, Henan, Shaanxi, and Ningxia, and parts of Gansu province in the west central part of the basin. The Huang He (Yellow River) watershed is identified as the cradle of Chinese civilization. It has supported an agricultural society for more than 7,000 years, longer than any other place on earth. It is also the home of the Han Chinese. This region has been the major cultural, economic, and political center of China for thousands of years, and it was the home of six ancient Chinese capitals. The southern boundary of this region is defined by the Qinling Mountains and Huai River geographic line. Beijing, as the present capital, is the home to administrative and architectural masterpieces, unsurpassed in the rest of the country.

The region has seen its share of floods, erosion, and earthquakes. Some people say this part of China has the longest continuous history of sorrow and vulnerability. The channel and waters of the Huang He shaped and molded the region. In the last 2,000 years, the river has changed its course about 10 times and flooded some 1,500 times. The river, as it winds its way through Gansu, Shaanxi, and Shanxi provinces, picks up the characteristic loess soil and displays the distinctive yellowish color for hundreds of miles as it finally flows into the Gulf of Bohai and the Yellow Sea. The problem of silting in the river is ongoing. Since silting, erratic water flow, and shifting shoals make navigation difficult, the Huang He has never functioned as a major river highway. Its use has been confined to local fishing, short trips, and local ferry and barge traffic. From early times, the Huang He was linked to the Chang Jiang by the manmade Grand Canal, the important artificial water channel for north-south transport of goods and people.

Shandong province, with the capital of Jinan and the seaport of Qingdao, comprises an open fertile plain through which the Huang He completes its journey to the sea. For centuries, the province was considered one of China's poorer regions. Yet the fertility of soil meant that human settlement here was continuous for more than 6,000 years. Qufu, located 100 miles (160 kilometers) directly south of Jinan, is the birthplace and burial site of Confucius. A revival of the province's economy is seen by a thriving tourist trade promoting sites such as Qufu and the former German port of Qingado. Just south of Jinan is the holiest of China's five sacred mountains, Taishan. A 5,070-foot (1,545-meter) climb leads to the summit that is one of the holiest places for Taoism. It also honors Buddhism and Confucianism. The Chinese people have probably worshipped at this site for longer than they have had written history. It has become a major pilgrimage site.

The famous Loess Plateau in the western part of the region is considered the ancient heartland of China. The loess deposits that cover the plateau are 200 feet (61 meters) deep, on average. Extensive erosion of the plateau and severe soil loss have plagued this area. Each year, up to 2 billion tons of soil wash into the river, and three-quarters of that reaches the Yellow Sea. The river is contained by dikes, built higher each year, to hold the constantly rising river channel. Traditionally, the Loess Plateau was a link connecting central China to the western part of the country via the Silk Road. Modern-day Xian, in the southwestern part of the Plateau, is home to the famous Terracotta Warriors. This ancient city has a distinctive perimeter wall that forms a rectangle whose sides total 8 miles (12 kilometers) in length.

The North China Plain is the largest and most populous expanse of flat, cultivated land in China. Villages in this region appear at regular intervals, and rows of trees mark fields and farms on the flat terrain. Diked canals appear in all directions. These are part of an ancient irrigation system, as well as an attempt to control the unpredictable Huang He.

Some researchers describe the North China Plain as an excellent example of a core area. It contains the capital and other major cities, including Tianjin, seaport on the Bohai Gulf. Tianjin also has a major industrial complex with heavy industry. In addition, the North China Plain is one of the most densely populated and highly productive agricultural areas in all of China.

Chang Jiang Drainage Area

Unlike the continuous, flat, agricultural and urban-industrial land associated with the sediment-laden Huang He to the north, the basins and valleys of the Chang Jiang Drainage Basin display variations in relief and elevation. The Chang Jiang begins its very long journey across the country in the mountains of northern Tibet. Some 700 tributaries contribute to its flow. Large cities have developed along its route—Chongqing, Wuhan, and Nanjing. Near its mouth lies one of China's largest cities, Shanghai. In the nineteenth century, several foreign countries saw Shanghai's potential as a harbor, and it developed as a banking, trading, and shipping center. It was the base of European imperialism. The river valley itself is estimated to occupy about one-fifth of China's land area, one-fourth of total farmland, and it holds one-third of China's population. Seventy percent of total rice production and 40 percent of total food production for China comes from the Chang Jiang Basin. The middle and lower portion of the river is one of the largest and busiest inland waterways in the world. The completion of the Three Gorges Dam will only make it busier.

The Grand Canal, the world's largest man-made waterway, dates back to 400 B.C. It linked the delta area of the Chang Jiang with the Huang He and Beijing to the north. The idea was to provide an easy route to carry any surplus rice from the lower Chang Jiang Basin to the populated parts of the north. In the past, the journey could take three months by barge.

Two important cities linked to the Grand Canal are Hangzhou and Nanjing. Hangzhou is the southern terminus, renowned for its silk and the scenic beauty of its famous West Lake. Marco Polo described this city near the end of the thirteenth century as the "City of Heaven." It is a wonderful environment of hills, trees, grass, pavilions, and pagodas. It was reported to be the favorite retreat of Mao, and the place where he entertained U.S. President Richard Nixon in 1972.

Nanjing, capital of Jiangsu Province, lies on the south bank of the Chang Jiang. It has been a capital at various times during the last 2,000 years. Its very name, "Southern Capital," stands in contrast to the "Northern Capital" of Beijing. Sun Yat-sen's impressive mausoleum is located in a beautiful park on the eastern edge of the city. One of the worst atrocities of World War II happened in Nanjing, when

Japanese soldiers slaughtered more than 300,000 civilians between 1937 and 1945.

Nanjing was walled as early as 2,500 years ago. Its city wall, 20 miles (32 kilometers) long, is the longest in the world. Part of the structure contains a colossal gate, which is actually four gates one inside another, making it the biggest of its kind in China. The magnificent three-mile-long (five kilometers) Yangtze River bridge is a source of Chinese engineering pride. Opened in 1968, it was the first major structure built by the Chinese after the Soviets left in 1960. It has a double-decker structure for train and vehicle crossings.

Upstream along the Chang Jiang, there is always an unending stream of large and small vessels, including "large trains" of six or more barges pushed by a tug. This is one of the busiest waterways in the world, and is the leading transit corridor for inland China. The completion of the Three Gorges Dam will further increase water traffic and improve navigation between coastal Shanghai and interior Chongqing.

The middle part of the Chang Jiang, south and east of Wuhan city, has a unique physical feature. This is China's best-known and extensive freshwater lake district, which includes an internationally recognized wetland zone. This series of rivers and lakes is essentially a network of small, shallow freshwater lakes and marshes fed by a multitude of rivers. Most are located in Hunan and Jiangxi provinces. The river network is quite dense, with many tributary systems. The entire area is quite extensive. Between 2,500 and 3,000 river systems can be identified. Alluvial plain development has allowed for great agricultural production, especially rice. Over the centuries, this lake district has acted as a reservoir for the flood waters of the Chang Jiang in summer. It is sometimes known as "the land of fish and rice." The climate is marked by rainy springs and hot summers, although frequent droughts can occur.

Two major lakes are Dongting Hu and Poyang Hu. Dongting Hu was once China's largest freshwater lake. This impressive, crescent-shaped lake covers 965 square miles (2500 square kilometers), and is fringed with reeds and lotus ponds and surrounded by farming villages. Very important for wildlife, Dongting Hu is home to the highly endangered Yangtzi river dolphin and Chinese sturgeon. Poyang Hu is a similar complex of small lakes and marsh areas, which can fluctuate seasonally. When summer floodwaters come, the lake district can reach 1,350 square miles (3,500 square kilometers), with the deepest point only 98 feet (30 meters). When the waters recede, the exposed land can support livestock and agriculture.

Sichuan Province, through which the Chang Jiang passes, is one of China's most populous. It is divided into two very different parts: a densely populated eastern plain and a mountainous west. The productive eastern part is the fertile Red Basin, named for its underlying red sandstone. The basin has been intensively cultivated for more than 2,000 years, and several kingdoms were founded here. It is said to contain the largest concentrated rice paddy cultivation in the world.

Xichang, 250 miles (400 kilometers) south of Chengdu, capital city of Sichuan province, is the site of China's space program. The Space Flight Center has seen the launch of communications satellites, with varying degrees of success. The city is on the same latitude—28° north latitude—as the Kennedy Space Center in Florida. Space launches closer to the equator require less fuel than those at higher latitudes.

Sichuan has always been a productive place, but problems loom on the horizon. Production must continue to keep pace with the population increase. There are also high levels of pollution produced by the basin's retention of warm moist air, combined with industrial and vehicle emissions. The maintenance of water quality standards for up to 100 million people remains an additional challenge.

Subtropical South China

This region comprises the area south of the Chang Jiang, including the provinces of Fujian, Yunnan, Guangdong, and Hainan Island and Guangxi AR. It also includes parts of Hunan, Jiangxi, and Zhejiang provinces, as well as the cities of Hong Kong and Macau. Since the seventeenth century, merchant classes prospered in this subtropical coastal environment. The entrepreneurial spirit emerged early and became very much a southern characteristic. Over time, the south thought of itself as separate from the rest of China, as demonstrated by the colonial enclaves of Hong Kong and Macau.

This region of China can be very green and lush, with a wet, steamy climate. It has rolling hills dotted with cultivated terraces for agriculture. Summers are hot, winters are mild, and rainfall is spread evenly throughout the year. Yet typhoons can plague this part of China. Rice and water buffalo are evident everywhere. Architecture is distinctive with two-story houses and an absence of walled courtyards. Language dialects are distinctive and complex because of differences in pronunciation and sentence structure. The indented coastline with its excellent harbors has fostered continuous foreign contact and trade dating back to the Tang Dynasty (618–907).

Traditionally, the southeast coast of China has been a window to the outside world. Fishing culture and access to open waters made contact with strangers from afar easy. Hong Kong and Macau have long been ports for this part of China. The Xi (West) River, China's southernmost river, forms a strategic delta area accompanying the Pearl River as both enter the South China Sea. The delta area is home to China's fastest-growing economic complexes, including Guangzhou (Canton) and Shenzhen, one of the fastest-growing cities in the world.

In south China, where the climate is subtropical, the humid and lush environment invites the use of traditional farming methods. There, the people sometimes use water buffalo as draft animals to help them work the extensive rice paddies.

Hong Kong is located on an irregularly shaped peninsula and a number of offshore islands beside the Pearl River delta to the west. Hong Kong includes the islands, the New Territories, and Kowloon. Formerly a British colonial outpost, since June 1997, Hong Kong has been a Special Administrative Region of

China. The city, once a cluster of small villages, is now a modern landscape with stylistic architecture and massive towering buildings that hug steep slopes. It is linked by an ultra-modern subway system to a new airport.

Forty miles from Hong Kong, Macau occupies the tip of a peninsula and a couple of offshore islands. It has evolved into a large casino playground. Its wealth can be seen in high-rise hotels, bridges, highways, and a new airport expansion. Its Portuguese colonial past pre-dates Hong Kong by 300 years, and some historic buildings and cultural features from the colonial era remain.

Capital of the province and a city of 5 million people, Guangzhou (Canton) is another vital component of this southeast region. Colonized by the Qin Dynasty in 221 B.C., it came under Chinese control during the Han Dynasty (206 B.C.–A.D. 220). Open to foreigners longer than any other city in China, it attracted foreign traders from as far away as central Asia and the Middle East. Guangdong Province is perhaps the most prosperous province in all of China, and certainly in south China. Located just south of the Tropic of Cancer, it has no real winter of which to speak. The area grows two crops of rice and a vegetable crop a year, as well as plenty of fruit.

Even though Guangzhou has long been the most important trading center in South China, the entire area has become one giant economic region. Today, there are five other trading and manufacturing centers along this south coast: Xiamen in Fujian Province; Shantou in eastern Guangdong Province; Shenzhen, adjacent to booming Hong Kong; Zhuhai across from Macau; and Haikou, on Hainan Island off Leizhou Peninsula.

The other distinctive part of subtropical China is the Yunnan-Guizhou Plateau. Noted for its natural beauty, the plateau is a rough, mountainous land. Yunnan Province is one of the most heavily forested regions in China. It is home to many of China's native plant species and one-third of China's 400 bird species. It has been called "The National Botanical Garden." The average elevation is 3,300 feet (1,005 meters) and reaches 6,500 feet (1,980 meters) on the northwest provincial border. The region has the most pleasant climate in China, warm and mild. The area is described as "spring at all seasons." It has a rich agriculture with cultivation of rice, winter wheat, tea, hemp, and beans. On the Vietnam-Laos border, the climate becomes hotter and wetter, and is ideal for growing rubber trees, sugar cane, as well as bananas and other tropical fruits and vegetables.

The region's outstanding natural wonder is Shilin, the Stone Forest, some 78 miles (125 kilometers) south of the capital city Kunming. It is really a 300-million-year-old exposed and extensive karst (limestone) formation of rocks weathered into unusual shapes. These high, jagged limestone columns that reach heights of 100 feet (30 meters) are interspersed with lakes. From a distance, the gray rocky outcrops resemble a petrified forest. There are many such "forests" in Yunnan, with the limestone rocks covered by extensive trees and vines.

Half of China's 55 minority peoples live in Yunnan. Guangxi-Zhuang Autonomous Region is home to the largest minority group, the Zhuang, with 20 million people. In contrast, the largest ethnic group, the Han, number more than 1.2 billion people.

The border areas between Sichuan, Yunnan, and Guizhou provinces are home to the Yi people, one of the largest ethnic minority groups in China. They number some 7 million people. Most of the Yi live in the mountainous areas, and for hundreds of years, they were isolated farmers. Thus, their language and shamanistic religion are unique in China. In shamanism, certain people are considered spiritually gifted, believed to possess the power to access good and evil spirits.

Inner Mongolia-Xinjian Steppelands

This is a vast extensive and remote area, long considered a peripheral part of China. This part of China, accounting for one-third of its land area, has often been viewed as territory lying beyond the Great Wall. For the Chinese, the region is considered remote, subject to the weather extremes, and occupied by "barbarians." This vast area of steppe and grassland, desert and mountain plateau, is home to significant minority peoples, such as Mongols, Uighurs, and Kazakhs. As such, the various parts are not considered provinces, but autonomous regions. The 6 million peoples of the Mongolian Steppelands and the 9 million Muslim Uighur peoples of the Xinjiang Autonomous Region of the northwest have been reluctant subjects of Chinese rule. Both areas have, since the eighteenth century, been brought under tight Chinese control. Daily life in this region is nomadic, quite different from other parts of eastern China.

China values this region not for its grasslands, but for what lies beneath. The region is considered the richest in the country for rare minerals. There are at least 60 verifiable mineral ores dispersed over 500 different locations. Such varieties include coal, iron, chromium, uranium, lead, zinc, gold, and salt.

The Xinjiang Autonomous Region is considered the least hospitable in all of China, covered by arid deserts and mountains. Its two giant basins are surrounded on all sides by mountains. To the north is the grassland Jungar Basin and to the south is the Tarim Basin, dominated by the hot and dry Taklamakan Desert. The region is more than 1,865 miles (3,000 kilometers) from any coast. It is the Tarim Basin where most of the Muslim Uighur peoples live. The Xinjiang Autonomous Region is usually associated with the Uighur people. Other minority people, such as the Kazakhs, Kyrgyz, and others, have cultural connections with the people across the border in the five republics of former Soviet Central Asia. A curious feature of this region is that the sun rises around 9 or 10 in the morning and sets at midnight, because China uses only one time zone, Beijing time.

Turpan, largely Uighur populated, is a fertile oasis situated in the Tian Shan. It is 505 feet (154 meters) below sea level, making it the second lowest point on the planet, second only to the Dead Sea in Jordan. Depressions such as these are a characteristic feature of this northern part of western China. The elevation difference here is 17,388 feet (5,300 meters), which creates a very hot and dry environment. Land temperatures can reach 104°F (40°C), with rainfall of only 16 millimeters per year. Famous for its grapes, the city of Turpan, called the "City of Grapes," is also known as the "hottest, lowest, driest, and sweetest" spot in China. It is also noted for its underground water channel, known as karez, or qanat. This is a system of shafts and irrigation channels that bring water melted from the nearby snow-capped mountains to the otherwise arid basin. This underground system, estimated at 2,000 miles (3,220 kilometers), prevents evaporation of water in the intense summer heat.

This remote region has extensive reserves of coal, oil, and natural gas. Experiments in space technology, rocketry, and sophisticated weaponry are conducted in this remote area, away from foreign observers. The isolated nature of this part of the country once gave the government a place to send notorious criminals and political opponents to prison camps.

Urumqi, the capital, is home to more than one million people. It is an important transportation hub in this northwest frontier, and was a strategic mountain pass route on the old Silk Road. It is inhabited by 13 minority nationalities, including Huis, Uighurs, Manchus, and Mongols. It is so vital to China as the most westerly industrial outpost that in 1992, it was officially declared a "port." Thus, the city received special low tax rates and other privileges usually permitted only in port cities such as Shanghai and Shantou.

China's most westerly city is Kashgar, with a population 300,000. It is located at 76°E longitude. It is over 2,485 miles (4,000 kilometers) from Beijing. This oasis city is the point at which the north and south arms of the Silk Road joined to form China's "front door," or natural gateway, through the Pamir mountain passes to Russia and India. The city was the spot where the so-called Great Game for Asia was played. Russian and British diplomats and spies converged here in the late 1800s to collect information on each other's activities. Today, it displays one of the few towering outdoor statues of Chairman Mao.

Tibetan Plateau

The "Roof of the World," called *Xizang* by the Chinese, is Tibet. The massive Tibetan Plateau, with an average elevation of 14,765 feet (4500 meters) above sea level, is surrounded on all sides by towering mountain ranges. It is the source area for at least five of the major rivers of east and south Asia. Some of the most spectacular scenery in the world is in this region of China, and human settlements appear to be dwarfed by high mountains. The Tibetan Plateau has an area of 965,255 square miles (2.5 million square kilometers), about one-fourth of China's total land area. This is the largest, highest, and geologically youngest plateau in the world. It contains the largest and most numerous lakes in China.

Tibet is the one part of China where Chinese troops have imposed harsh rule on the population and, in the process, forced the political and spiritual leader, the fourteenth Dalai Lama, to flee the country. Since the early 1950s, Tibet has been subjugated by the Chinese government, which encouraged Han migrations into Lhasa, the Tibetan capital region, and along the India border. The Chinese administer Xizang as an Autonomous Region.

In the remote region of the Tibetan Plateau, many people still practice the lifestyles they knew long before the creation of China's modern government. Some, like this man, even remain nomadic, wandering the area in search of food, without a permanent home.

Distinctive Tibetan Buddhist culture began in the eighth century and was flourishing by the eleventh century. The traditional economy is based on raising and grazing animals, including sheep, goats, horses, and cattle. It is the yak, a draft animal, however, that best symbolizes the culture of this area. The yak provides meat, milk, cheese, butter, hide, fur, hair, butterfat, and dung that is used for fuel.

Women have always played an important and self-sufficient role in the society of this region. Because of the nomadic lifestyle, marriage in Tibet was complex. Women were free to have more than one husband, and men were free to have more than one wife, in a system called polygamy. This is not the prevailing social custom, but it is more common in Tibet than elsewhere.

Unique in this region of China is the great variation in temperature. Diurnal range, the difference between daytime and nighttime temperatures, can reach 82°F (28°C). July temperatures in Tibet are the lowest in China, at 50°F (10°C).

Despite decades of Chinese control, and the sometimes brutal repression inflicted on Tibet, the area's traditions continue, the Buddhist monasteries still function, and the beauty of the landscape remains. The Tibetan Plateau and its people remain an enigma for the central government of China.

The great progress China has made toward economic stability and improvement in human rights issues is evident in the selection of the nation's capital, Beijing, as the site for the 2008 Summer Olympic Games. When the International Olympic Committee announced Beijing's selection in July 2001, the people took to the streets in a jubilant celebration.

China Looks Ahead

The twentieth century saw dynamic forces change the political landscape of China. Shortly after the century began, the ages-old imperial dynastic system was swept away. By mid-century, China was ruled by a Communist system under the authoritarian Chairman Mao Zedong. By century's end, China was forced by global realities to experiment with forms of capitalism. Along the way, the country survived occupation by foreign troops, a short-lived republican government, a failed attempt at monarchical restoration, a war against Japan, and five years of civil war.

The People's Republic of China emerged on October 1, 1949. Over the last half century, there has been major land and social reform, the Great Leap Forward Campaign, the Cultural Revolution, famine, the market reforms of Deng Xiaoping, and the move to put China on the world political and economic map. Through it all,

Chinese civilization has endured and even thrived. The rich history and cultural heritage has equipped its people well to forge ahead. There is a strong sense of unity that has held the Chinese nation together through 4,000 years. Its stable territorial boundaries, borders that defy easy penetration, culture, language, rich philosophy, and political institutions have allowed China to remain unified through dynastic changes and periodic social upheavals. The country has proved resilient and enduring. The future should prove no different.

Decades of reform have created many vested interests, both collective and private, in the growth of the Chinese economy. Such growth is now linked to the global economy. China is a huge country with enormous resources, both natural and human, and has much untapped potential. There are vast reserves of coal and oil. The Three Gorges Dam symbolizes the massive hydroelectric potential of the country. Some observers feel that China should have the world's largest economy within just a few decades. Yet, this booming economy brings with it the potential for further environmental degradation. Even today, air and water availability and quality are deteriorating, which demands immediate action. For China, the unprecedented threat for future survival comes not from outside but from within. Never has Chinese civilization faced such immense internal problems. This is a challenge that will only get worse if not addressed.

Some people argue that democracy may not be suitable for China. The nation's long history with dynastic rule and centralized authority, along with its huge territory and growing population, may prevent full democracy in the foreseeable future. As in the dynasties of the past, a strong military seems to be essential to allow the central government to exercise social control and curb the threat posed by political demonstrations, such as the one that occurred in Tiananmen Square in 1989. Three million soldiers and 1.2 million reserves help exert internal control over society. The military of China will continue to have an integral role to play in the twenty-first century. On

The military has long been a vital part of China's government and culture, and has always helped maintain the strength of the nation's central government. Although some nations today fear the expanding power of the Chinese military, most observers believe the armed forces are necessary for the stability of the nation.

paper, the country has the largest standing military in the world, but it has an antiquated arsenal and poorly equipped troops. However, it does possess atomic weapons and nuclear technology. It is developing a navy. It continues to enhance its foreign intelligence-gathering capabilities. In the first half of the 1990s, China was the sixth largest arms exporter and seventh largest arms importer in the world. The Chinese military will continue to be watched by the Western powers.

The status of the offshore island of Taiwan touches a raw nerve in Beijing, since the Chinese regard the island as a province of China. The development of more democratic institutions in Taiwan acts as a counter to the authoritarianism and one-party domination of the mainland government. The future of relations between China and Taiwan remains uncertain. The two systems seem likely to coexist for some time, with occasional disagreements, but the two economies will most likely become ever more intertwined.

The immediate future for China's economy, as in the past, will continue to rely on agriculture. Food production must keep pace with the population increase. The one-child policy should help bring population growth under control. The projected population change for China between 2001 and 2050 is 8 percent, compared with 58 percent in India, where similar birth control measures have not been implemented. Famines have occurred in recent times. Such extreme conditions, both natural and human, might bring social unrest and foster regional uprisings that could challenge the central government. Deng Xiaoping introduced the innovative idea that "it is glorious to be rich." If such riches are not distributed among all the Chinese people and shared by the entire country, however, China could face serious internal problems.

Over the centuries, the Chinese have always had a strong attachment to the land. Family and land were the cornerstones on which Chinese society rested. Increasing personal mobility

China's economic growth and rise to power among the major countries of the world has been very impressive, and the nation will most likely remain a significant player in international affairs in the years to come.

and the trend toward a technological-industrial rather than agrarian society could unravel the traditional Chinese social fabric. China's past has been one of slow and steady adaptation to change. It is perhaps with this strategy, which governed the nation's past greatness, that China's best approach to an uncertain future lies.

Facts at a Glance

Country Name	People's Republic of China (PRC) Zhonghua Renmin Gonghe Guo (Chinese)
Location	Eastern Asia, bordering the Korea Bay, Yellow Sea, East China Sea, Taiwan Strait, the South China Sea, and the Gulf of Tonkin, between Hainan Island and North Vietnam.
Capital	Beijing
Flag	The red background represents the Communist revolution. The large star is a symbol for the Communist Party. The four smaller stars represent the workers, peasants, petty bourgeoisie (capitalists), and national bourgeoisie who are united in building a new society.
Land Area	3.7 million square miles (9.6 million square kilometers)
Coastline	9,010 miles (14,500 kilometers)
Land Boundaries	13,760 miles (22,147 kilometers)
Border Countries	North Korea 880 miles (1,416 kilometers), Russia (northeast) 2,240 miles (3,605 kilometers), Mongolia (2,905 miles (4,677 kilometers), Russia (northwest) 25 miles (40 kilometers), Kazakhstan 950 miles (1,533 kilometers), Kyrgyzstan 530 miles (858 kilometers), Tajikistan 260 miles (414 kilometers), Afghanistan 45 miles (76 kilometers), Pakistan 325 miles (523 kilometers), India 2,100 miles (3,380 kilometers), Nepal 770 miles (1,236 kilometers), Bhutan 290 miles (470 kilometers), Burma 1,360 miles (2,185 kilometers), Laos 260 miles (423 kilometers), Vietnam 795 miles (1,281 kilometers).
Climate	Extremely diverse; tropical in south to cold subarctic in the north; eastern coast dominated by seasonal reversal of winds called Asiatic monsoon.
Terrain	Plains, deltas, hills in east; mountains, high plateaus, deserts in west.
Highest Point	Mount Everest (Qomolangma) 29,035 feet (8,850 meters)
Lowest Point	Turpan Pendi (Depression) -505 feet (-154 meters)
Land Use	Arable: 10%, includes irrigated at 5% Meadows and pastures: 43% Forests and woodland: 14% Other: 33%

Natural Hazards	Frequent typhoons (hurricanes)–up to seven per year along southern and eastern coasts; floods, earthquakes, droughts, infrequent tsunamis.
Environment–Current Issues	Air pollution (greenhouse gases, sulfur dioxide particulates) from heavy use of coal produces acid rain; water shortage, particularly in the north; water pollution from untreated wastes; deforestation; soil erosion; desertification; loss of plant and animal species.
Administrative Divisions	22 provinces (Taiwan is considered the 23rd) 5 autonomous regions (ARs) 4 directly administrated municipalities 2 special administrative regions (SARs)
Monetary Unit	Renminbi, RMB (Yuan)
Labor Force by Occupation	Agriculture and forestry 50% Services 28% Industry 22%
Industries	Iron and steel, coal, machine building, armaments, textiles and apparel, petroleum, cement, chemical fertilizers, footwear and clothing, toys, food processing, automobiles, consumer electronics, telecommunications.
Primary Imports	United States $397.4 billion (2003 est.) Machinery and equipment, mineral fuels, chemicals, plastics, iron and steel
Import Partners	Japan 18%, Taiwan 11.9%, South Korea 10.4%, United States 8.2%, Germany 5.9% (2003 est.)
Primary Exports	United States $436.1 billion (2003 est.) Machinery and equipment, textiles and clothing, footwear, toys and sporting goods, mineral fuels
Export Partners	United States 21.1%, Hong Kong 17.4%, Japan 13.6%, South Korea 4.6%, Germany 4% (2003 est.)
Population	1.3 billion
Population Density	332 people per square mile (128 per square kilometer)
Life Expectancy at Birth	71.96 years Female: 73.72 Male: 70.40 (2004 est.)

Facts at a Glance

Age Structure	0–14 years 22.3% 15–64 years 70.3% Over 64 years 7.5%
Ethnic Groups	Han Chinese 91.9%; others 8.1%, including Zhuang, Manchu, Miao Hui, Uighur, Yi, Tujia, Tibetan, Mongol.
Religions	Officially atheist, but PRC recognizes four religions: 6% Buddhist, 2% Taoist, 2% Muslim, 1% Christian.
Languages	Standard Chinese or Mandarin (Putonghua, based on Beijing dialect), Yue (Cantonese), Wu (Shanghaiese), Min (Fukienese).
Literacy	90.9% (2002)
Government Type	Communist/Socialist republic
Head of State	President
Independence	221 B.C. (unified under Qin Dynasty). Qing Dynasty replaced by Republic February 12, 1912. People's Republic of China (PRC) established October 1, 1949.
Transportation	Highways: 870,000 miles (1.4 million kilometers) Railways: 35,000 miles (68,000 kilometers) Airports: 206 (2000 est.) Waterways: 68,350 miles (110,000 kilometers)

History at a Glance

600,000–400,000 B.C.	First hominids–Lantian Man and Peking Man
80,000 B.C.	Appearance of modern man, *Homo sapiens*, in China
7,000 B.C.	Beginnings of agriculture and of Neolithic period
5,000 B.C.	Yangshao culture Longshan culture
2100–1600 B.C.	Xia–earliest recorded dynasty
1600–1027 B.C.	Shang Dynasty flourishes in Huang He (Yellow River) Valley
1027–771 B.C.	Western Zhou Dynasty First mathematical textbooks Mandate from Heaven concept originated
770–256 B.C.	Eastern Zhou Spring and Autumn Period (722–476 B.C.) Confucius (551–479 B.C.) Laozi, founder of Taoism (570–490 B.C.) Warring States period (475–221 B.C.)
221–206 B.C.	Qin Dynasty First emperor unites China Great Wall completed Terracotta army statues built to guard tomb at Xian
206 B.C.–220 A.D.	Han Dynasty Confucianism accepted as state ideology Western Han (206 B.C.–A.D. 24) Eastern Han (A.D. 24–220) Buddhism enters China Technology of paper-making Seismograph invented
220 A.D.–581	Disunity and Partition Known as "three kingdoms and six dynasties" Chinese economic center shifts south to Chang Jiang (Yangtze River)
581–618	Sui Dynasty Grand Canal built from Hangzhou to Beijing Great Wall rebuilt
618–907	Tang Dynasty China's Golden Age; arts and literature flourish

History at a Glance

907–960	Five Dynasties and Ten Kingdoms Invasion of nomadic tribes
1023	Chinese are first to use paper currency
960–1279	Song Dynasty Northern Song Dynasty (960–1127) Southern Song Dynasty (1127–1279) Marco Polo in China (1271–1292)
1279–1368	Yuan Dynasty (Mongol rule) Kublai Khan (1214–1294)
1368–1644	Ming Dynasty Sea voyages to South China Sea, the Indian Ocean, and Africa First Dalai Lama (1447) in Tibet Arrival of Western traders
1644–1911	Qing Dynasty (Manchu rule) Opium Wars (1839–1842) Hong Kong ceded to Great Britain (1842) Taiping Rebellion (1850–1864) Boxer Uprising (1900)
1877	Former U.S. President Ulysses S. Grant visits China
1899	Future U.S. President Herbert Hoover works in China as a mining engineer
1911	End of 2,000 years of imperial rule in China Sun Yat-sen proclaimed president of Republic of China
1921	Communist Party founded in Shanghai
1926	Jiang Jieshi becomes Nationalist leader after Sun Yat-sen dies
1927	Nationalists purge Communists
1931	Japan seizes Manchuria
1934–1935	Mao and followers retreat to northwest China on Long March
1937	Japan invades China; Nationalists and Communists form United Front
1937–1945	China and Japan at war
1941	Japan bombs Pearl Harbor; United States becomes ally of China U.S. volunteer fliers form "Flying Tigers" in air bases in China outside of Japanese occupation
1945	Japan surrenders
1946–1949	Civil war ends as Nationalist forces are defeated by Communist

	People's Liberation Army (PLA)
1949	Jiang Jieshi and Nationalists flee to Taiwan and Mao Zedong proclaims People's Republic of China
1950	Korean War begins; China sends troops into Korea
1951	China sends PLA units to Tibet
1953	Korean War ends in a truce
1956–1957	"Hundred Flowers" campaign; critics of government are later punished
1958	"Great Leap Forward" campaign and People's Communes established; widespread famine
1959	Tibetan uprising brings harsh reprisals from Beijing; Dalai Lama flees to India
1960	Rift between China and Soviet Union
1962	China in border war with India
1964	China explodes first nuclear bomb; thirty more bombs follow over the years
1966–1976	Cultural Revolution
1969	Chinese clash with Soviet troops at Ussuri River border in northeast Manchuria
1971	People's Republic of China replaces Taiwan at the United Nations
1972	Richard Nixon, becomes first sitting U.S. president to visit China
1975	Jiang Jieshi dies; U.S. President Gerald Ford visits China
1976	Mao Zedong dies; Gang of Four arrested
1977	Deng Xiaoping takes charge in China
1979	United States recognizes PRC Deng visits United States 16-day border war between China and Vietnam
1981	Special Economic Zones created along east coast
1984	U.S. President Ronald Reagan visits China
1989	Brutal suppression of democracy movement in Tiananmen Square in Beijing. Jiang Zemin chosen as party secretary-general U.S. President George H.W. Bush visits China

1990	Asian Games in Beijing
1997	Deng Xiaoping dies; Hong Kong returns to Chinese rule; Chinese President Jiang Zemin visits United States
1998	U.S. President Bill Clinton visits China
1999	Portugal returns Macau to China
2001	China joins World Trade Organization; Summer Olympics awarded to Beijing for 2008
	U.S. President George W. Bush visits Shanghai for Asia Pacific Economic Conference
2002	U.S. President George W. Bush makes first official visit to China
	CCP Sixteenth Congress meets
2003	Hu Jintao becomes party general-secretary replacing Jiang Zemin. SARS outbreak originated in Guangdong Province. Three Gorges Dam sluice gates close to allow reservoir to fill up. First Chinese taikonaut (astronaut) Lt. Col. Yang Liwei, orbits earth 14 times in 21 hours.
2004	Jiang Zemin steps down as chairman of the Military Commission.

alluvial soil: Material deposited by rivers and streams, usually rich and fertile.

Bodhisattva: Buddhist "saints" who have attained enlightenment, but remain on earth to help others.

Boxers: Participants in an anti-foreign uprising among peasants that first began in Shandong province in 1898. Formally known as the "Society of Righteous and Harmonious Fists;" followers of an ancient art of self-defense and heterodox beliefs.

double-cropping: The planting and harvesting of two crops a year on the same plot of land.

Gang of Four: Four people, including Mao's wife, accused of persecuting up to a million innocent people during the Cultural Revolution (1966–1976). After Mao's death, they were arrested, tried, and condemned in 1980.

Han Dynasty: An important and powerful dynasty (206 B.C.–A.D. 220) of China that created the first large-scale empire in East Asia. It came to be used as the name of the majority ethnic group.

karez: An extensive underground irrigation system to carry water by gravity from nearby mountains to arid flatlands below, in desert zones such as western China. Sometimes known as qanat.

karst: Limestone rock where water erosion creates a landscape with sinkholes, caverns, underground streams, and residual hills (haystack hills).

loess: A very fine silt or sand deposited by wind over a considerable distance. When irrigated, it is very fertile. It can stand in steep vertical walls.

Long March: In 1934–1935, the Chinese Communists survived a trek of 6,000 miles (9,660 kilometers) from southern Jiangxi province west and then north to Yenan in the loess hills of Shaanxi province.

Middle Kingdom: The traditional Chinese view of their country as the center of the known universe.

monsoon: A climate system affecting coastal Asia based on seasonal changes in pressure and wind patterns. In the hot summer months, wet monsoons blow onshore with heavy rainfall. Dry monsoons blow offshore in the cool season.

People's Communes: Artificially organized rural communities following the Communist ideology, they came into existence in 1958 together with China's Great Leap Forward program.

PLA: The People's Liberation Army, the official name of Chinese military forces since the founding of the Chinese People's Republic in 1949.

Qing Dynasty: The last imperial dynasty (1644–1911) of China, established by the Manchu conquest of China.

Special Economic Zones (SEZ): Six zones created in the 1980, along China's southern and east coast, designated by the government and given special tax and investment privileges to attract foreign investment and technology.

steppe: Semi-arid open grassland, sometimes synonymous with prairie. Climate usually has hot summers, cold winters, and low annual rainfall.

subsistence farming: The growing of crops for direct use by farmers and families, with emphasis on self-sufficiency.

Three Gorges Dam: When completed in 2009, it will be the largest hydroelectric dam in the world. Situated on the middle portion of Chang Jiang (Yangtze) River.

tsunami: A very large sea wave created by an underwater seismic jolt (earthquake or volcanic eruption), capable of causing extensive coastal damage. Frequently misnamed a "tidal wave."

typhoon: A fierce tropical storm created by a deep, circular low pressure system, generated by heat and water. In the Western Hemisphere, it is called a hurricane.

Yuan: China's unit of legal tender, known officially as Renminbi.

Avedon, John F. *In Exile From the Land of Snows.* New York: Alfred A. Knopf, 1984.

Benewick, Robert, and Stephanie Donald. *The State of China Atlas.* New York: Penguin, 1999.

Cannon, Terry, and Alan Jenkins, eds. *The Geography of Contemporary China: The Impact of Deng Xiaoping's Decade.* New York: Routledge, 1990.

Cotterell, Arthur. *China: A Cultural History.* New York: Penguin, 1988.

Cressey, George. *Asia's Land and Peoples: A Geography of One-Third of the Earth and Two-Thirds of Its People,* 3rd Rev. ed. New York: McGraw-Hill, 1963.

Dreyer, June Teufel. *China's Political System: Modernization and Tradition,* 3rd ed. London: Longman, 2000.

Economy, Elizabeth C. *The River Runs Black: The Environmental Challenge to China's Future.* Ithaca, New York: Cornell University Press, 2004.

Fairbank, John K., and Merle Goldman. *China: A New History.* Cambridge: Belknap Press of Harvard University Press, 1999.

Haw, Stephen G. *A Traveller's History of China,* 3rd ed. New York: Interlink Books, 2001.

Hersey, John. *A Single Pebble.* New York: Alfred A. Knopf, Inc., 1956.

Ho, Yong. *China. An Illustrated History.* New York: Hippocrene Books, Inc., 2000.

Hook, Brian, ed. *The Cambridge Encyclopedia of China,* 2nd Rev. ed. New York: Cambridge University Press, 1994.

Hunter, Alan, and John Sexton. *Contemporary China.* New York: St. Martin's Press, 1999.

Lardy, Nicholas R. *Agriculture in China's Modern Economic Development.* Cambridge: Cambridge University Press, 1983.

Leeming, F. *The Changing Geography of China.* Oxford: Blackwell, 1993.

Mackerras, Colin. *China's Minorities.* Hong Kong: Oxford University Press, 1994

Mackerras, Colin and Yorke, Amanda. *The Cambridge Handbook of Contemporary China.* Cambridge: Cambridge University Press, 1991.

Menzies, Gavin. *1421.* London: Bantam Books, 2003.

Ryder, Grainne. *Damming the Three Gorges: What Dam-Builders Don't Want You to Know.* Toronto: Probe International, 1990.

Salisbury, Harrison E. *The Long March: The Untold Story.* New York: Harper and Row, 1985.

Further Reading

Shambaugh, D., ed. *Greater China: The Next Superpower?* Oxford: Oxford University Press, 1995.

Snow, E. *Red Star Over China.* London: Gollancz, 1968.

Spence, Jonathan D. *The Chan's Great Continent: China in Western Minds.* New York: W.W. Norton, 1998.

Tsai, Jung-fang. *Hong Kong in Chinese History.* New York: Columbia University Press, 1993.

Tuan, Yi-Fu. *The World's Landscapes, China.* London: Longman, 1970.

Winchester, Simon. *The River at the Centre of the World: A Journey Up the Yangtze and Back in Chinese Time.* London: Viking, 1997.

Wong, Jan. *Red China Blues: From Mao to Now.* New York: Doubleday and Co., 1999.

Zhao, Songqiao. *Geography of China: Environment, Resources, Population and Development.* New York: Wiley, 1994.

Internet Addresses

Creaders Net News
http://www.creadersnet.com/

China News Digest
http://www.cnd.org

Chinese Embassy in Washington
http://www.china-embassy.org

Far Eastern Economic Review
http://www.feer.com/

Human Rights Watch/Asia
http://www.hrw.org

Tibet Information Network
http://www.tibetinfo.net

Xinhua New Agency
http://www.xinhua.org

Index

Index

Picture Credits

page:

10: Reprinted with permission from *Current History* magazine (September 2002) © 2002 Current History, Inc.
11: CIA/University of Texas, Austin
12: Keren Su/CORBIS
17: AP/Wide World Photos
20: Michael S. Yamashita/CORBIS
28: Kevin R. Morris/CORBIS
39: Keren Su/Corbis
42: Quadrillion/CORBIS
47: Stapleton Collection/CORBIS
55: AP/Wide World Photos
58: AP/Wide World Photos
67: Kevin R. Morris/CORBIS
71: Macduff Everton/CORBIS
72: AP/Wide World Photos
79: AP/Wide World Photos
81: AP/Wide World Photos
84: Tom Nebbia/CORBIS
90: Julia Waterlow; Eye Ubiquitous/CORBIS
95: AP/Wide World Photos
104: Liu Liqun/CORBIS
114: The Purcell Team/CORBIS
120: Michael S. Yamashita/CORBIS
122: AP/Wide World Photos
125: Wolfgang Kaehler/CORBIS
127: Keren Su/CORBIS

Frontis: Flag courtesy of *theodora.com/flags*. Used with permission.

About the Author

GARY T. WHITEFORD, a Canadian, earned degrees in geography, a BA from York University in Toronto, an MA from Clark University, and a PhD from the University of Oklahoma. He has been teaching geography in the Faculty of Education at the University of New Brunswick since 1974. He has presented over 20 papers at various conferences over the years, authored numerous journal articles, and co-edited an acclaimed atlas of the world. The author has lived in Japan and travelled extensively in China and parts of Southeast Asia. Dr. Whiteford is an active member of the National Council for Geographic Education and the Canadian Association of Geographers.

CHARLES F. "FRITZ" GRITZNER is Distinguished Professor of Geography at South Dakota State University. He is now in his fifth decade of college teaching and research. Much of his career work has focused on geographic education. Fritz has served as both president and executive director of the National Council for Geographic Education and has received the Council's George J. Miller Award for Distinguished Service.

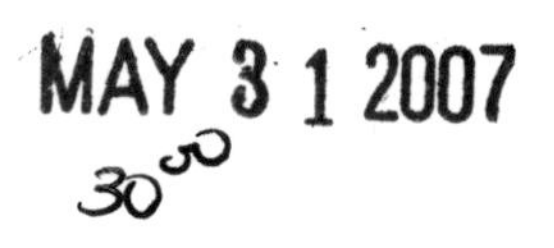
MAY 3 1 2007